Petrography of Igneous and Metamorphic Rocks

Petrography of Igneous and Metamorphic Rocks

Anthony R. Philpotts
The University of Connecticut

CBS Publishers & Distributors Pvt. Ltd.
New Delhi • Bengaluru • Chennai • Kochi • Mumbai • Pune
Hyderabad • Kolkata • Nagpur • Patna • Vijayawada

WAVELAND PRESS, INC.
Long Grove, Illinois

Petrography of Igneous and Metamorphic Rocks

Waveland ISBN: 978-1-57766-295-2

Reissued 2003 by Waveland Press, Inc.

The worldwide edition of this book is published by Waveland Press, Inc., 4180, IL, Route 83, Suite 101, Long Grove, Illinois 60047, United States of America

CBS Reprint: 2015

CBS ISBN: 978-81-239-2665-0

Published by:
Satish Kumar Jain for CBS Publishers & Distributors Pvt. Ltd.,
4819/XI Prahlad Street, 24 Ansari Road, Daryaganj, New Delhi - 110002
delhi@cbspd.com, cbspubs@airtelmail.in • www.cbspd.com
Ph.: 23289259, 23266861, 23266867 • Fax: 011-23243014

Corporate Office: 204 FIE, Industrial Area, Patparganj, Delhi - 110 092
Ph: 49344934 • Fax: 011-49344935
E-mail: publishing@cbspd.com • publicity@cbspd.com

Branches:

- ***Bengaluru:*** 2975, 17th Cross, K.R. Road, Bansankari 2nd Stage, Bengaluru - 70 Ph: +91-80-26771678/79 • Fax: +91-80-26771680
  E-mail: cbsbng@gmail.com, bangalore@cbspd.com
- ***Chennai:*** No. 7, Subbaraya Street, Shenoy Nagar, Chennai - 600030
  Ph: +91-44-26681266, 26680620 • Fax: +91-44-42032115
  E-mail: chennai@cbspd.com
- ***Kochi:*** 36/14, Kalluvilakam, Lissie Hospital Road, Kochi - 682018
  Ph: +91-484-4059061-65 • Fax: +91-484-4059065
  E-mail: cochin@cbspd.com
- ***Mumbai:*** 83-C, Dr. E. Moses Road, Worli, Mumbai - 400018
  Ph: +91-9833017933, 022-24902340/41 • E-mail: mumbai@cbspd.com
- ***Pune:*** Bhuruk Prestige, Sr. No. 52/12/2+1+3/2,
  Narhe, Haveli (Near Katraj-Dehu Road Bypass), Pune - 411041
  Ph: +91-20-64704058/59, 32342277 • E-mail: pune@cbspd.com

Representatives:

- Hyderabad: 0-9885175004
- Kolkata: 0-9831437309, 0-9051152362
- Nagpur: 0-9021734563
- Patna: 0-9334159340
- Vijayawada: 0-9000660880

Printed at: India Binding House, Noida, U.P.

To my Parents

Contents

Preface

Petrography of Igneous and Metamorphic Rocks is designed for students taking their first course in petrography. It combines, in a single book, a compilation of the optical properties of the common rock-forming minerals, descriptions of the textures and structures of igneous and metamorphic rocks, and a treatment of the classification of these rocks. The book deals with the description of rocks, not their origins. The coverage is not exhaustive and is limited to what students are likely to encounter and need to know in an introductory petrography course. The material is organized so as to make it as rapidly accessible as possible. For those who have not previously had a course in optical mineralogy, Chapter 2 briefly introduces this subject.

The minerals described are the so-called rock-forming minerals. These are the major building constituents of rocks, with which the student of petrography must have some familiarity. For quick reference, they have been listed alphabetically in Chapter 3. Some minerals, such as the amphiboles, micas, pyroxenes, and feldspars, have been listed together under their group names. The most important distinguishing properties of these minerals are compiled into a single table on the inside front cover. For routine petrography, this table provides most of the information necessary to distinguish the common minerals. The interference colors shown in this table are those found in 30-μm-thick sections, the standard thickness of a petrographic thin section.

The classification of igneous rocks used in this book is the one proposed by the International Union of Geological Sciences (IUGS) Subcommission on the Systematics of Igneous Rocks. In addition, a list of commonly used rock names, many of which are not part of the IUGS classification, is keyed to this classification. Also, the widely used Irvine-Baragar classification of volcanic rocks is included.

Illustrations of minerals and rock textures are grouped together at the end of appropriate chapters. Each of these 216 illustrations is shown in color on the CD included with the book. The CD provides a convenient means for the beginning petrographer to gain extra experience examining thin sections.

ACKNOWLEDGMENTS

In a book such as this, in which many data are compiled, all mistakes are difficult to eliminate. However, careful reviews of the manuscript by Steven R. Bohlen, State University of New York at Stony Brook, Christopher I. Chalokwu, Auburn University, Susan C. Eriksson, Virginia Polytechnic Institute and State University, Timothy M. Lutz, University of Pennsylvania, and Stearns Anthony Morse, University of Massachusetts, Amherst, have helped eliminate most. The ones that remain are the author's responsibility, and he hopes that users of the book will communicate any errors that they find to him. The author is particularly grateful to his colleagues Richard P. Bennett, A. J. (Mike) Frueh, Norman H. Gray, Raymond L. Joesten, and Randolph P. Steinen who have helped in various ways during the preparation of the manuscript. The greatest debt of gratitude, however, is owed the many students who, through their questions over the years, have kept the author peering down microscopes.

Anthony R. Philpotts

1 Introduction

Petrography is the science of describing and classifying rocks. It relies heavily on observations made with the petrographic microscope, but observations on the outcrop and with the hand lens are also important. A petrographic description of a rock first involves identification of the minerals and, where possible, determination of their compositions. Textural relations between grains are noted, for these not only help in classification but provide evidence of processes active during the formation of the rock. The rock is then classified on the basis of the volume percentages of the various rock-forming minerals (mode).

In an introductory petrography course a student can become familiar with only a small fraction of the great variety of rocks found in nature. Fortunately, the number of important types is surprisingly small. This is because rocks are formed in only a few tectonic environments on Earth, and the conditions in these have changed little, if any, throughout most of geologic time. Thus a Precambrian basalt appears identical to a modern one. A few rocks, however, are peculiar to the early geologic record.

Although the goal of petrography is the description and classification of rocks, this in itself is of limited interest. Only when considered as part of petrology (the study of the origins of rocks) does it take on wider significance. Petrography supplies most of the data that petrology strives to interpret and explain. It is expected, therefore, that students will take a separate petrology course. It is desirable, but not essential, that some practice in optical mineralogy has previously been gained.

A number of excellent books provide all of the information needed for petrography (see suggested readings at the end of this chapter). These include reference works on the optical properties of minerals (Deer et al., 1962-63, 1966; Fleischer et al., 1984; Ramdohr, 1969; Troger, 1979; Cameron, 1961), atlases of colored photographs of minerals and rocks (MacKenzie et al., 1982), and petrography texts (Williams et al., 1982; Spry, 1969; Harker, 1954; Nockolds et al., 1978). These books provide far more information than can be used in an introductory course. Furthermore, a selection of books including one from each of these groups involves a considerable expense. Students should, however, be aware of these sources of information and make use of them for extra reading.

This laboratory manual is a collection of the essential information needed to learn the techniques of petrography and to apply them to common igneous and metamorphic rocks. It presents the optical properties of all the minerals likely to be encountered in the common rocks. It also describes the textures and classifications of igneous and metamorphic rocks. Sedimentary rocks are not included, because the manual is designed for the laboratory part of a course in igneous and metamorphic petrology. The optical properties of the common sedimentary minerals are, nonetheless, included in the manual.

The manual contains a large number of figures illustrating the rock-forming minerals and the textures of rocks. The figures are black and white drawings rather than photographs, for these allow emphasis to be given to the properties being illustrated. Color, especially that due to birefringence, is an important optical property of a mineral, but inclusion of color photographs would substantially increase the cost of the manual. The figures have therefore been reproduced in color on the CD that is bound with the book.

The figures in the text and the CD are in four groups covering Igneous Rock-Forming Minerals (IRFM), Metamorphic Rock-Forming Minerals (MRFM), Igneous Textures (IT), and Metamorphic Textures (MT). Figures in each group are numbered consecutively from one, and references to them in the manual are given by the reference letters followed by the number of the figure. Beside each figure is a short caption and sufficient space to make your own notes on the features that you feel are important in identifying a mineral or in characterizing a rock texture.

Proficiency in petrography comes only with practice and experience. It is desirable to devote more time to studying petrography than is available in the normally scheduled laboratory periods in a course. It is recommended therefore that, in addition to the work done with the microscope in the laboratory, time be spent studying the figures in this manual (including the CD) or in the texts referenced below.

RECOMMENDED READINGS

Optical Mineralogy

Bloss, F. D., 1961, An Introduction to the Methods of Optical Mineralogy, Holt, Rinehart and Winston, New York, 294 p.

Shelley, D., 1985, Optical Mineralogy, Elsevier, New York, 321 p.

Stoiber, R. E., and Morse, S. A., 1972, Microscopic Identification of Crystals, Ronald Press, New York, 278 p. (1981 edition, Krieger, Melbourne, Florida)

Wahlstrom, E. E., 1979, Optical Crystallography, John Wiley Sons, New York, 488 p.

Optical Properties of Minerals

Cameron, E. N., 1961, Ore Microscopy, John Wiley Sons, New York, 293 p.

Deer, W. A., Howie, R. A., and Zussman, J., 1962 and 1963, Rock Forming Minerals, 5 vols., Longman, London.

Deer, W. A., Howie, R. A., and Zussman, J., 1966, An Introduction to the Rock Forming Minerals, Longman, London, 528 p.

Fleischer, M., Wilcox, R. E., and Matzko, J. J., 1984, Microscopic Identification of the Nonopaque Minerals, U.S. Geological Survey Bull. 1627.

MacKenzie, W. S., and Guilford, C., 1980, Atlas of Rock-Forming Minerals in Thin Section, John Wiley Sons, New York, 98 p.

Phillips, W. R., and Griffen, D. T., 1981, Optical Mineralogy: The Nonopaque Minerals, W. H. Freeman and Company, San Francisco, 677 p.

Ramdohr, P., 1969, The Ore Minerals and Their Intergrowths, (English translation of the 3rd ed.), Pergamon Press, Oxford, 1174 p.

Troger, W. E., 1979, Optical Determination of Rock-Forming Minerals, E. Schweizerbart'sche Verlagsbuchhanlung, Stuttgart, 4th ed (English edition by Bambauer, H. U., Taborsky, F., and Trochim, H. D.) 188 p.

Petrography

Harker, A., 1954, Petrology for Students, Cambridge University Press, London, 8th ed. (revised by C. E. Tilley, S. R. Nockolds, and M. Black), 283 p.

Johanssen, A., 1931-1939, A Descriptive Petrography of the Igneous Rocks, 4 vols., University of Chicago Press, Chicago.

MacKenzie, W. S., Donaldson, C. H., and Guilford, C., 1982, Atlas of Igneous Rocks and their Textures, John Wiley Sons, New York, 148 p.

Moorhouse, W. W., 1959, The Study of Rocks in Thin Section, Harper Row, New York, 514 p.

Nockolds, S. R., Knox, R. W., and Chinner, G. A., 1978, Petrology for Students, Cambridge University Press, London, 435 p.

Spry, A., 1969, Metamorphic Textures, Pergamon Press, Oxford, 350 p.

Williams, H., Turner, F. J., and Gilbert, C. M., 1982, Petrography: An Introduction to the Study of Rocks in Thin Sections, W. H. Freeman and Company, San Francisco, 626 p.

2 Review of the Microscope and Mineral Optics

The Microscope

Most petrographic microscopes are of essentially the same basic design. They are expensive, precision instruments and should be treated with considerable care. The user must be familiar with their workings and be aware of any problems that may arise. Avoid touching the glass in the lenses, and at no time should any moving part of the microscope be forced; instead ask for assistance. For purposes of illustration, the Carl Zeiss standard polarizing microscope is shown in Fig. 2-1.

Modern microscopes are fitted with a built-in light source. On some instruments the brightness of this light is adjustable. For general work, the light intensity need not exceed three-quarters of the maximum value. With normal tungsten bulbs, a blue filter must be placed over the light source in order to render the light similar to daylight. If this is not done, the colors of minerals under the microscope will be distinctly yellow.

The eyepiece, or ocular, normally has a magnification of 8X, 10X, or 12.5X. Multiplication of this factor by the magnification of the objective lens gives the total magnification of the microscope. The eyepiece contains cross hairs which must be focussed by each individual user. In doing this, take care not to position the eye too close, or too far from the lens, for this diminishes the field of view (Fig. 2-1). Relax the eye so that it is focussed at infinity; this can be done by momentarily looking at a distant object. Then, keeping both eyes open, bring the cross hairs into sharp focus by rotating the adjustable eyepiece. You may find this can be performed more easily by removing the eyepiece from the microscope. It can then be placed in front of the eye while you observe some distant object. When replacing the eyepiece, note that keyed slots on the side of the microscope tube allow the cross hairs to be oriented in the 90° (N-S, E-W) or 45° (N-W, N-E) positions. The N-S, E-W orientation is used most commonly.

Most microscopes are equipped with three objective lenses--low, intermediate, and high magnifications--and possibly a fourth, an oil immersion lens for extra high magnifications. Each lens is marked with a series of numbers, such as 40/0.85-160/0.17, which indicates the objective has a magnification of 40X and a numerical aperture of 0.85 in a microscope with a tube length of 160 mm, and it is to be used with microscope slides covered with 0.17 mm-thick cover glasses. The total magnification of the microscope (ie., objective x eyepiece) cannot exceed 1000X the numerical aperture. Magnifications above this are said to be empty; an enlarged image is obtained, but with no gain in resolution. Very low magnifications can be obtained with the Bertrand lens and no objective at all. The image produced is, however, of poor quality.

Objective lenses are commonly mounted on a turret that can be rotated by a knurled wheel; **never** rotate the turret by pushing on the lenses; this will almost certainly cause misalignment of the lenses. Each of the lenses must be centered with respect to the axis of rotation of the microscope stage. The axis of rotation can be identified as that point on a microscope slide which does not move when the stage is rotated. By use of centering rings or screws on the objective lens, this point is moved till it coincides with the image of the cross hairs.

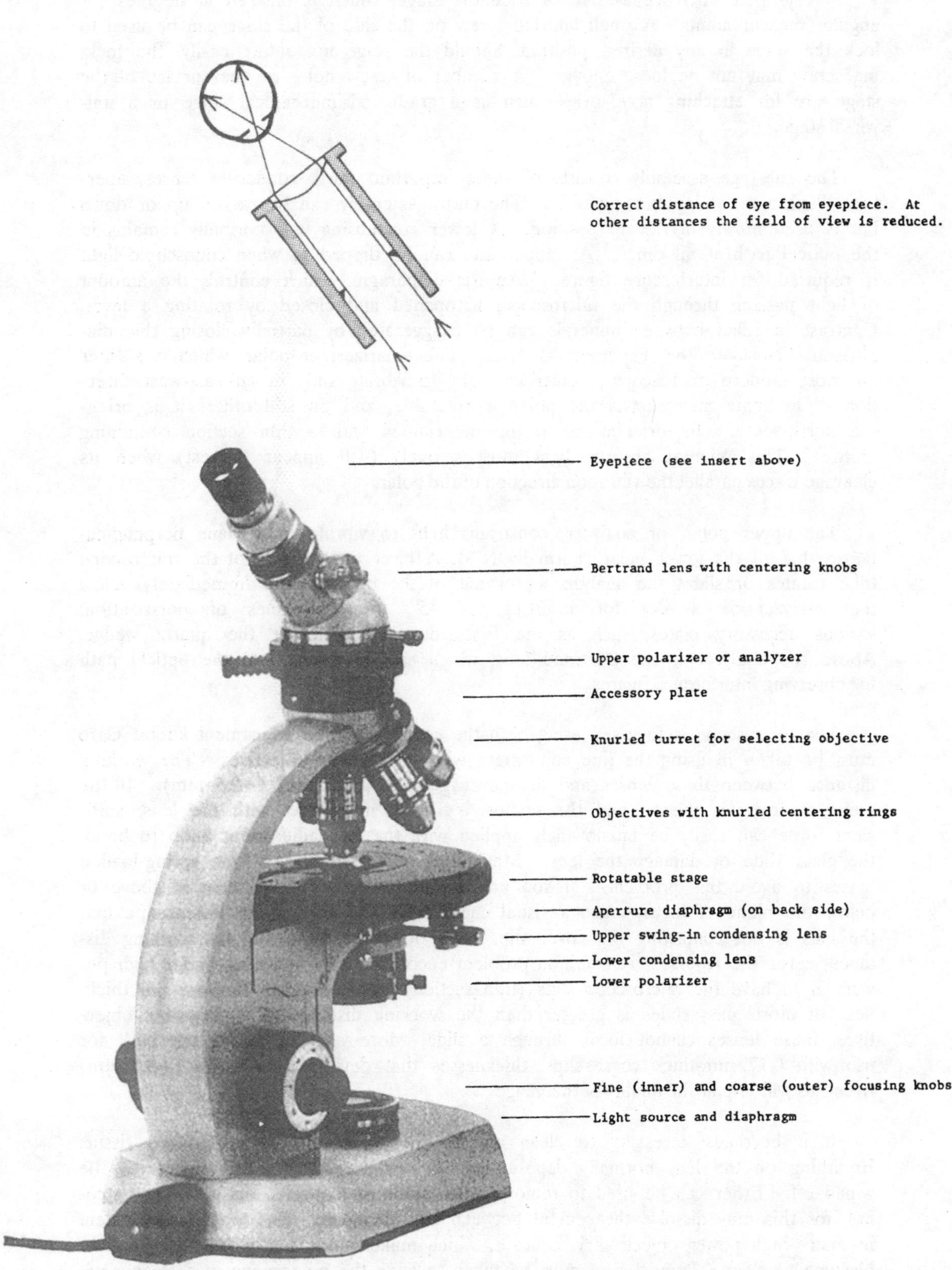

Figure 2-1 Carl Zeiss Standard Polarizing Microscope (reproduced with permission of Carl Zeiss, West Germany).

Petrographic microscopes have a rotatable stage, which is marked in degrees for angular measurements. A small knurled screw on the side of the stage can be used to lock the stage in any desired position. Should the stage not rotate easily, the locking screw may not be loose enough. A number of screw holes on the surface of the stage are for attaching accessories, such as a graduated mechanical stage or a universal stage.

The substage assembly consists of three important parts--condenser lenses, aperture diaphragm, and polarizing filter. The entire assembly can be racked up or down but is used mostly in the up position. A lower condensing lens normally remains in the optical path at all times. An upper one can be flipped in when conoscopic light is required for interference figures. An iris diaphragm, which controls the amount of light passing through the microscope, is opened and closed by rotating a lever. Contrast in relief between minerals can be exaggerated by partially closing this diaphragm. Normally the diaphragm is open. The polarizer, or polar, which is a filter on most modern microscopes, constrains light to vibrate only in an east-west direction. On some microscopes the polar is rotatable, and on still others it is oriented north-south. Its orientation can be determined with a thin section containing biotite. This mineral absorbs light most strongly (will appear darkest) when its cleavage traces parallel the vibration direction of the polar.

The upper polar, or analyzer, constrains light to vibrate in a plane perpendicular to that of the lower polar (normally N-S). A lever on the side of the microscope tube rotates or slides the analyzer in or out of the optical path. Immediately below the analyzer is a slot for inserting, at 45° to the planes of polarization, various accessory plates, such as the first-order red plate or the quartz wedge. Above the analyzer is the Bertrand lens, which can be swung into the optical path for observing interference figures.

The microscope is focussed using both the coarse and fine adjustment knobs. Care must be taken in using the fine adjustment with high power objectives. The working distance between these lenses and the cover glass is very small (~0.6 mm). If the point of focus is missed and the section is raised in contact with the lens, sufficient force can easily be unknowingly applied with the fine adjustment knob to break the glass slide or damage the lens. Many high power objectives have spring-loaded lenses to avoid this problem. If you are doubtful whether the section is above or below the plane of focus, make a visual check from the side of the microscope that the lens is not contacting the cover slip. Then, focus by increasing the working distance--never the reverse. A common problem encountered in focussing under high powers is to have the microscope slide (thin section) upside down. Because the thickness of most glass slides is greater than the working distance of high power objectives, these lenses cannot focus through a slide. Moreover, the lenses are built for use with 0.17 mm-thick cover slips; thicknesses that deviate significantly ($\pm$0.03mm) from this will impair the quality of the image.

If it becomes necessary to clean any of the lenses, do so using lens tissue. Breathing on the lens normally deposits enough water so that finger prints can be wiped off. Ether can be used to remove more stubborn deposits, but never use alcohol, for this may dissolve the cement between lens elements. The front lens element in many high power objectives is concave, which makes these lenses particularly troublesome to clean. Utmost care must be taken to keep the microscope in a clean environment at all times, and a dust cover should be used when the instrument is not in use.

Passage of Light through a Crystal

The velocity of light passing through most crystalline substances depends on the direction of propagation relative to the crystal structure. Only in crystals belonging to the isometric system and amorphous materials, such as glass, is the velocity the same in all directions; such material is said to be *isotropic*. Other crystalline substances are said to be *anisotropic*; that is, the velocity of light varies with direction in the crystal. Although variations in the velocity of light are responsible for most of the optical properties used to identify minerals, the velocity of light itself is not determined. Instead, we measure the refractive index, which is the ratio of the velocity of light in air to that in the mineral; that is,

Refractive Index (R.I.) = velocity in air / velocity in mineral.

From this relation, the refractive index is seen to be inversely related to the velocity of light in the mineral. Thus a mineral with a high refractive index will transmit light more slowly than one with a low refractive index.

When light enters an anisotropic crystal it is constrained to vibrate in two mutually perpendicular planes, the orientations of which depend on the symmetry and orientation of the crystal. The petrographic microscope uses polarized light in order to determine the orientation of these vibration directions and to investigate other optical effects associated with this phenomenon.

Light emanating from the microscope lamp, on passing through the lower polar is constrained to vibrate in only one direction, which in most microscopes is E-W (Fig. 2-2). The light rising from the polar can therefore be represented by a sinusoidal electromagnetic wave that vibrates in an EW plane. This light, which is said to be plane polarized, is cut out completely by the upper polar, which passes light vibrating only in a NS direction. If, however, we could rotate the upper polar to a NE-SW direction, as can be done on research microscopes, the plane-polarized light rising from the lower polar would transmit a component of its vibration through the upper polar, but the amplitude of the transmitted wave would be less than that of the original wave (Fig. 2-2).

Plane-polarized light, on entering an anisotropic crystal, will resolve itself into two components that vibrate in the mutually perpendicular planes. The light vibrating in these planes will not, except in special cases, have the same velocity of propagation. There are one or two directions in anisotropic crystals along which the velocities of the two components are identical. These special directions are known as *optic axes*. Crystals having only one optic axis are known as *uniaxial*, and those with two, as *biaxial*. Uniaxial crystals belong to the trigonal, hexagonal, or tetragonal symmetry systems, whereas the biaxial ones belong to the orthorhombic, monoclinic, or triclinic systems.

Because the velocity of light vibrating in different directions in anisotropic crystals varies with direction, so must the refractive index. In uniaxial crystals, only a maximum and minimum refractive index need be defined, but in biaxial ones, an additional intermediate value must be given. In any anisotropic crystal the difference between the maximum and minimum refractive indices is known as the *birefringence*.

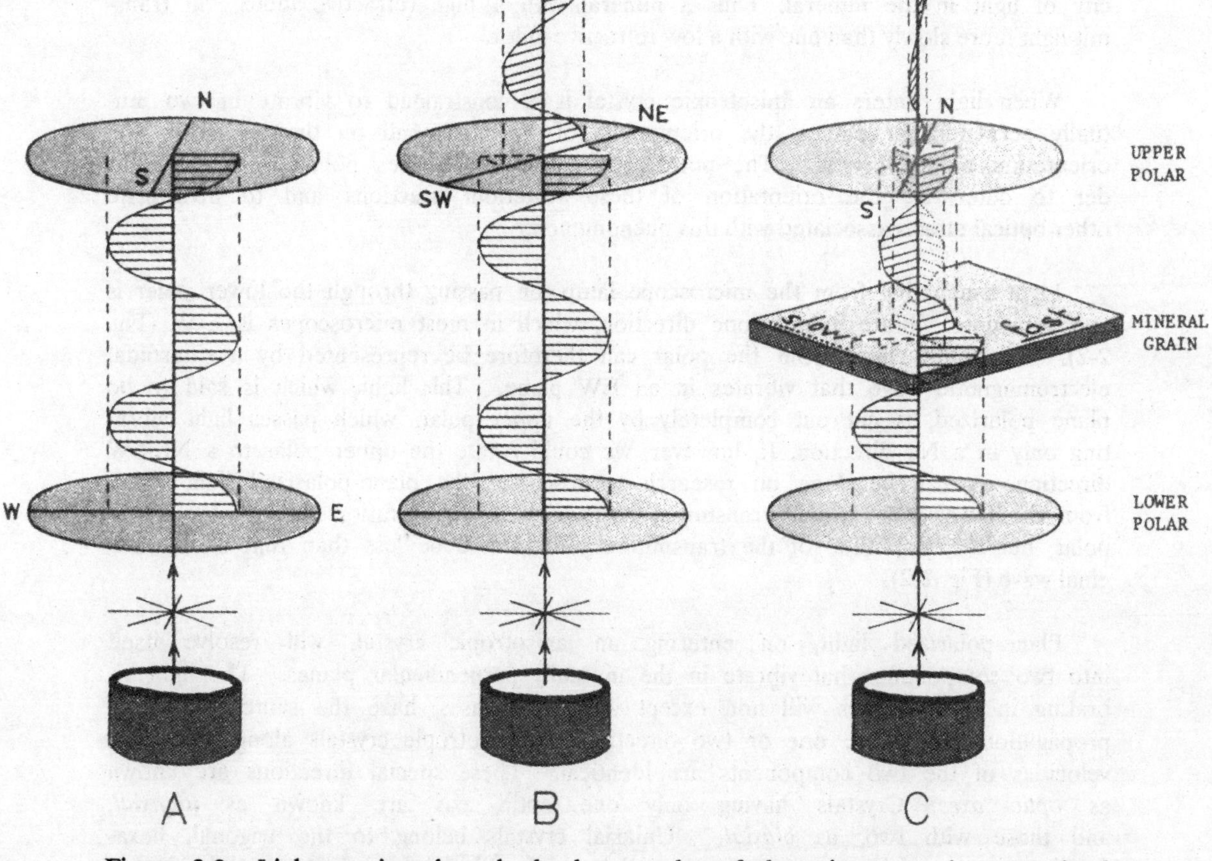

Figure 2-2 Light passing through the lower polar of the microscope is constrained to vibrate only in an E-W direction. In A, the upper polar, which allows light to vibrate only in a N-S direction, cuts out completely the E-W polarized light, and no light passes through the microscope. In B, the upper polar has been rotated to a 45° position, so that a component (diminished in amplitude) of the E-W polarized light is transmitted. In C, the plane polarized light is constrained to vibrate in two mutually perpendicular planes in an anisotropic mineral. One of these directions is the fast and the other the slow vibration direction. Both vibration directions produce a component in the N-S plane of the upper polar, but because of the different velocities of transmission, the two waves are out of phase when recombined by the upper polar, and an interference color results.

When an anisotropic mineral is placed on the stage of the microscope, the plane-polarized light from the lower polar is resolved into the two vibration directions permitted by that crystal; these two directions are then resolved into one by the upper polar (Fig. 2-2). Because the light in the two vibration directions travels with different velocities, the waves are out of phase when combined by the upper polar, and an interference color results. The amount by which the waves are out of phase, known as the *retardation*, depends on the difference in the velocities (hence in the refractive indices) of the two vibration directions and the thickness of the crystal; that is,

Retardation = Thickness x Difference in Refractive Indices.

Because thin sections are normally ground to a thickness of 30 μm, variations in retardation result primarily from variations in refractive index differences (which depend on crystal orientation). Small grains that do not extend through the entire thickness of a section will, however, have smaller retardations because of reduced thickness.

A retardation of 100 nm produces a gray interference color, whereas 300 nm produces a yellow color (see inside front cover). Colors resulting from retardations between 0 and 550 nm are said to be *first order.* First order red changes to second order blue as the retardation increases above 550 nm. Second and higher order colors repeat at retardations which are multiples of 550 nm, but the higher the order the less intense is the color. For example, first order red is deep red, whereas third order red is pinkish.

If the vibration directions in an anisotropic crystal are rotated so as to parallel the polars, no light is transmitted through the microscope, and the mineral is said to be in *extinction.* The orientation of extinction positions with respect to identifiable crystallographic directions, such as crystal outline, twin planes, or cleavage planes, provides a useful diagnostic property.

In addition to extinction positions it is useful to identify which of the two mutually perpendicular vibration directions is the faster. This is done by rotating the grain to the 45° position and inserting an accessory plate, such as the first order red one, in the slot provided. The fast and slow vibration directions in the accessory plate are marked on the plate. Should these correspond with the fast and slow directions in the mineral, their combined effect will be to increase the retardation and produce a higher order color. On the other hand, if the fast vibration direction in the plate parallels the slow direction in the mineral, the combined retardation will be less, and a lower order color will result. The fast, intermediate, and slow directions in a biaxial crystal are labeled x, y, and z respectively.

For most minerals the first order red plate is a convenient accessory for determining the fast and slow vibration directions. The retardation in highly birefringent minerals, however, may give such high order colors that the addition or subtraction of first order red may not produce a noticeable color change. For such cases, the quartz wedge can be used. As the wedge is pushed into the accessory slot, an increasingly thicker section of quartz is introduced into the optical path. In this way the amount of retardation is increased until it produces a color change that is discernible. The quartz wedge can also be used to determine the birefringence of minerals by rotating the grain into the 45° position that aligns the fast direction in the mineral with the slow direction in the wedge. As the wedge is slowly inserted, the order of the colors will decrease until the grain goes black. Then, slowly removing the wedge, count the order of colors as they increase to the initial color. The birefringence can then be determined from the color chart on the inside front cover.

Light passing from one material to another of different refractive index is refracted by an amount given by *Snell's law*:

$$\frac{\text{R.I. } 2}{\text{R.I. } 1} = \frac{\sin i}{\sin r}$$

where the various terms are defined in Fig. 2-3. When a ray of light passes from a medium of lower to one of higher refractive index, the ray is refracted towards the normal to the interface. Thus, with grains having a higher refractive index than their surroundings, light tends to be concentrated above the grain (Fig. 2-3). If a grain has a lower refractive index than the surroundings, the light tends to concentrate over the surroundings. By focussing above the grain (increasing the working

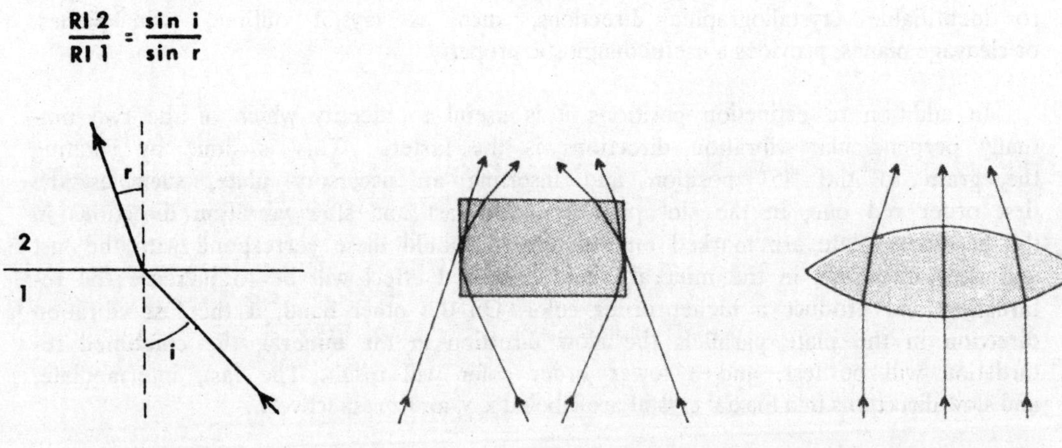

Figure 2-3 Light passing from a medium of higher refractive index to one of lower refractive index (R.I. 2 > R.I. 1) is refracted towards the normal to the interface according to Snell's law, where i is the angle of incidence and r the angle of refraction of the light ray. Refraction of light by mineral grains that have a higher refractive index than that of their surroundings results in light being concentrated above the grain, as seen for convergent and parallel light and for two different-shaped grains. If the refractive index of the mineral were less than that of the surroundings, the light would be concentrated over the surrounding material.

distance) a zone of brightness, or *Becke line*, will be seen to move into the material of higher refractive index. In thin sections, this allows the refractive index of grains to be determined relative to that of their surrounding grains or mounting medium. The term *relief* refers to the difference in refractive index between the mineral and the mounting medium. Prior to the use of epoxies, Canada balsam, with a refractive index of 1.537, was the main medium used. Thus, minerals with a higher refractive index than 1.537 have positive relief, whereas those with lower, have negative relief.

The color a mineral has in thin section under plane-polarized light (upper polarizer removed) depends on the color of the light source (normally white) and the wavelengths of light absorbed by the mineral. Not all wavelengths that constitute white light are necessarily absorbed to the same extent by a mineral. Preferential absorption of colors at the red end of the spectrum gives a mineral a bluish color in transmitted light, whereas absorption of blue light makes the mineral appear red. Furthermore, anisotropic minerals may absorb differently in different directions, with the result that they change color when rotated in plane polarized light. This gives rise to the property known as *pleochroism*.

The Optical Indicatrix

The variation in refractive index with direction in anisotropic crystals is conveniently represented by a geometrical figure known as an optical *indicatrix* (Fig. 2-4). It is an ellipsoid in which the vector from the origin to the surface gives the magnitude of the refractive index for light vibrating in that particular direction. Within the crystal, this light is constrained to vibrate in two mutually perpendicular planes, the traces of which are the maximum and minimum axes of the elliptical section through the indicatrix normal to the direction of light propagation; that is, the plane of the thin section (Fig. 2-5). Except for the special directions of the optic axes, these two vibration directions will have unequal velocities and thus different refractive indices.

In uniaxial crystals, that is, ones with only one optic axis, the indicatrix is an ellipsoid of revolution. The axis of revolution is the vibration direction of the so-called extra-ordinary ray (e). The refractive index for light vibrating in this direction is commonly designated by the Greek letter epsilon (ϵ). The circular section of the ellipsoid is the vibration direction of the ordinary ray (o). Light propagating perpendicular to this section can vibrate in any radial direction and it will have the same velocity regardless of vibration direction. This propagation direction is known as the *optic axis*. The refractive index for the ordinary ray is commonly designated by the letter omega (ω). If the refractive index for the extraordinary ray is greater than that for the ordinary ray ($\epsilon > \omega$), the crystal is said to have a *positive* sign, whereas if $\epsilon < \omega$ it is *negative* (Fig. 2-4).

The indicatrix for a biaxial crystal (two optic axes) is a triaxial ellipsoid in which the refractive indices associated with the fast (x), intermediate (y), and slow (z) vibration directions are the magnitudes of the three principal axes (Fig. 2-4). The refractive indices in these directions are referred to, respectively, as alpha (α), beta (β), and gamma (γ). Two circular sections exist in a triaxial ellipsoid. The refractive index for light propagating normal to either of these is equal to that of the intermediate value (β). The normals to the circular sections are the *optic axes*, and the angle between them is the *optic angle*, or 2V. The crys-

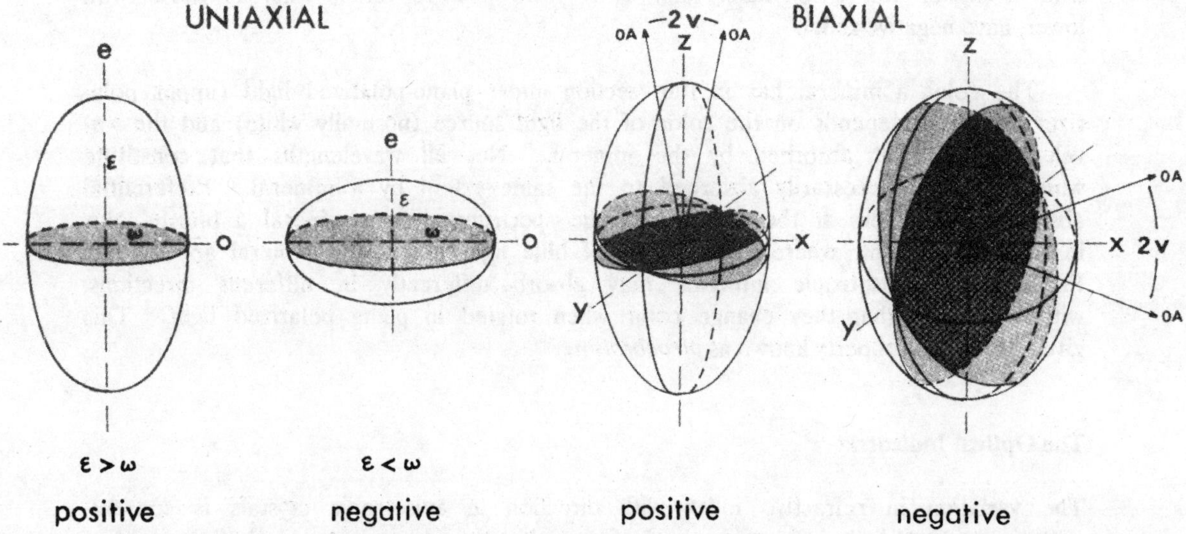

Figure 2-4 The optical indicatrix is a graphical means of representing the magnitude of the refractive index (distance from origin to surface of ellipsoid) for light vibrating in that particular direction in an aniso- tropic mineral. In uniaxial minerals, the indicatrix is an ellipsoid of revolution whose equatorial plane is a circular section (shaded) through the ellipsoid. Light travelling normal to this plane has the same velocity (equal refractive index) regardless of vibration direction; this unique direction is the optic axis. The biaxial indicatrix is a triaxial ellipsoid which has two circular sections through it (shaded) and two optic axes normal to these. The acute angle between the optic axes is known as 2V. The positive and negative sign conventions are indicated.

tal is said to be *positive* if the slow vibration direction, z, lies in the acute optic angle, and *negative* if the fast direction, x, occupies this position. In common usage, the acute angle is referred to as the optic angle. But often it is convenient, especially with minerals where 2V increases beyond 90° with variation in composition, to refer the optic angle to a particular axis, for example, $2V_x$.

Plane-polarized light entering a randomly oriented crystal on the stage of the microscope will be resolved into two mutually perpendicular vibration directions. One of these directions in a uniaxial crystal must be the ordinary vibration direction (Fig. 2-5). The extra-ordinary direction, however, will not be the maximum value unless the indicatrix is lying in the unique position with the **e** direction on the stage of the microscope. In a general position, therefore, this vibration direction is designated **e'**. A randomly oriented biaxial crystal may have none of the principal axes of the indicatrix lying in the plane of the section (Fig. 2-5); so the vibration directions are referred to as **x'**, **y'**, and **z'** to indicate the axes they most closely approach.

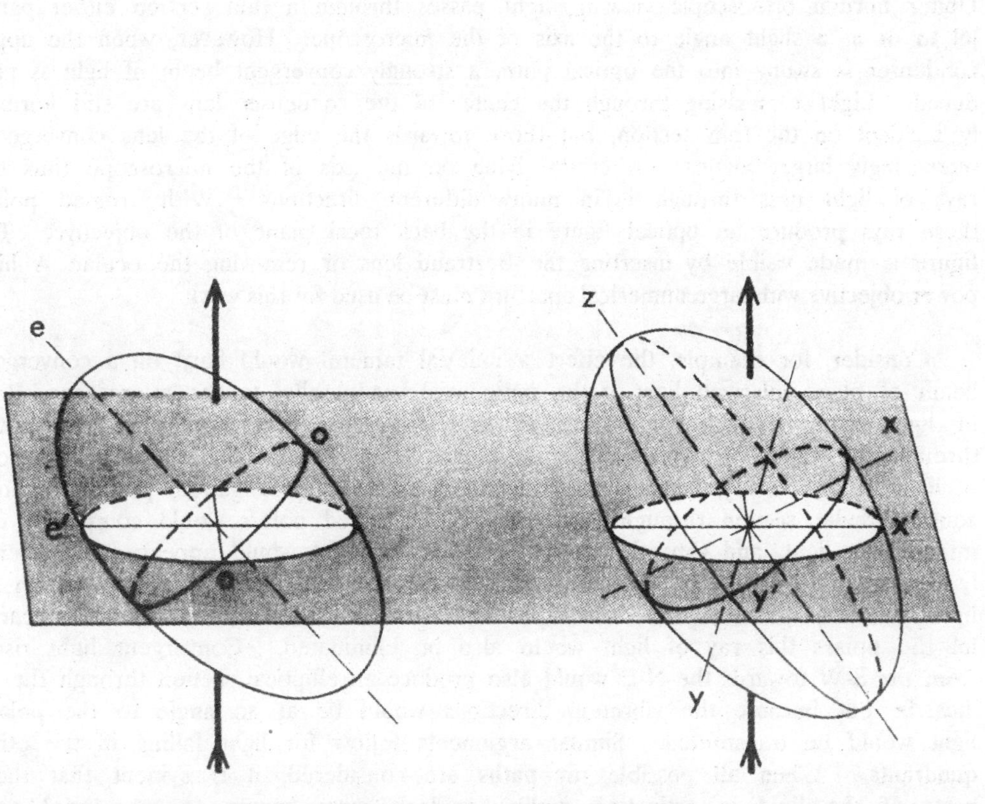

Figure 2-5 Plane-polarized light passing vertically through a randomly oriented grain in a thin section (shaded plane) will be constrained to vibrate in two mutually perpendicular directions. In a uniaxial crystal, one of these directions always corresponds to the ordinary vibration direction (o), but the extraordinary direction in most orientations will be less than the maximum possible and is therefore designated by e'. In a biaxial crystal neither of the vibration directions need correspond to the x, y, or z vibrations; thus they are designated by any two of x', y', or z', depending on which axes of the indicatrix are closest to the vibration directions.

Because the indicatrix is an expression of the interaction of light with the crystal structure of a mineral, the orientation of the indicatrix must obey the restrictions imposed by the symmetry of the crystal. Thus, because all uniaxial crystals belong to either the trigonal, hexagonal, or tetragonal systems, the optic axis must be the c crystallographic axis. The symmetry of biaxial crystals, likewise, restricts the orientation of the indicatrix. A mirror plane, for example, forces one of the principal axes of the indicatrix to be normal to the mirror plane. Only in the triclinic system is the indicatrix free to take on any orientation.

Refractive indices listed in this manual are referred to the vibration directions with which they correspond (e and o, or x, y, and z) rather than using the Greek letters (ϵ and ω, or α, β, and γ). This is done because of the convenience of listing in tables other properties which can also be related to these directions.

Observation of Crystals under Conoscopic Light

Under normal orthoscopic viewing, light passes through a thin section either parallel to or at a slight angle to the axis of the microscope. However, when the upper condenser is swung into the optical path, a strongly convergent beam of light is produced. Light rays rising through the center of the condenser lens are still normally incident on the thin section, but those towards the edge of the lens converge at increasingly larger angles. A crystal lying on the axis of the microscope thus has rays of light pass through it in many different directions. With crossed polars these rays produce an optical figure in the back focal plane of the objective. The figure is made visible by inserting the Bertrand lens or removing the ocular. A high power objective with large numerical aperture must be used for this work.

Consider, for example, the effect a uniaxial mineral would have on a convergent beam of plane-polarized light if the optic axis was parallel to the microscope. Rays of light rising through the center of the convergent lens would be transmitted through the crystal parallel to the optic axis (Fig. 2-6). The vibration directions of light in such a ray, being normal to the direction of propagation, produce a horizontal circular section through the indicatrix. Crossed polars would completely eliminate this light, and thus the central part of the field would appear dark. Strongly convergent light rising from the west towards the east would give rise to an elliptical section through the indicatrix, but because the vibration directions parallel the polars this ray of light would also be eliminated. Convergent light rising from the S-W towards the N-E would also produce an elliptical section through the indicatrix, but because the vibration directions would be at an angle to the polars, light would be transmitted. Similar arguments follow for light falling in the other quadrants. When all possible ray paths are considered, it is evident that those parts of the field in extinction outline a dark cross known as a *uniaxial optic axis figure*. The dark bands of this figure are known as *isogyres*.

The color of light transmitted in each of the quadrants of a uniaxial optic axis figure depends on the birefringence of the mineral and the angle of convergence of light, which varies radially with the distance from the center of the figure. Because light towards the edge of the field of view converges at a greater angle than that near the center, it passes through the crystal at a greater angle to the optic axis. As a result, the more convergent light exhibits greater birefringence. Concentric isochromatic rings may surround the optic axis if the birefringence of the mineral is sufficiently great.

The sign of a uniaxial mineral can be determined easily from an optic axis figure. In Fig. 2-6 it will be seen that every section through the indicatrix formed by the various ray paths contains the ordinary vibration direction. At right angles to this and following radial lines is a component of the extra-ordinary ray, e'. When the first-order red accessory plate is inserted, the isogyres turn red and the interference colors in adjacent quadrants increase or decrease depending on whether the vibration directions in these quadrants and in the accessory plate match or oppose one another. The fast vibration direction in most accessory plates is parallel to its length. If the mineral is positive (R.I. $\epsilon > \omega$), as is the case in this example, light vibrating parallel to the extra-ordinary ray will be slower than that parallel to the ordinary ray. In the N-W and S-E quadrants the retardation between these rays will be opposite to that in the accessory plate and the interference colors will decrease, for example, to yellow. In the N-E and S-W quadrants, however, the retardations match, and the interference colors increase, for example, to second

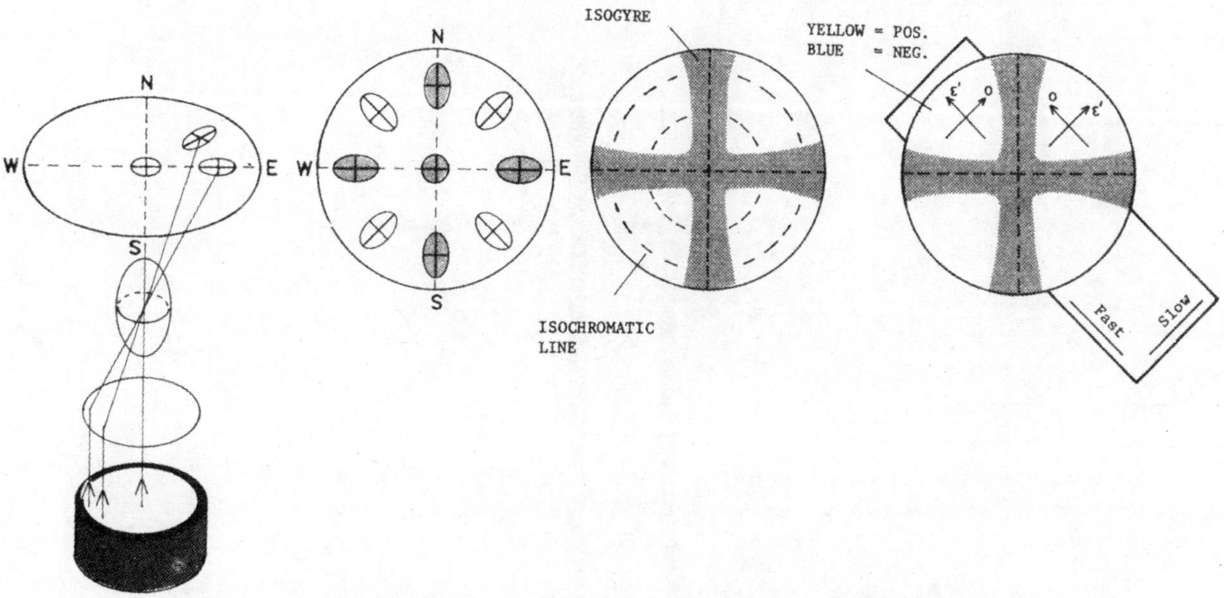

Figure 2-6 Generation of a uniaxial optic axis figure. Conoscopic light passing through a uniaxial mineral with its optic axis oriented parallel to the axis of the microscope will vibrate in the two directions indicated by the principal axes of the elliptical sections through the indicatrix. Where these vibration directions parallel the polarization planes of the polars, extinction occurs; where they do not, light is transmitted. The color of the transmitted light depends on the birefringence of the mineral; also the interference colors increase outward and may produce concentric isochromatic rings. Insertion of a first-order red accessory plate causes isogyres to become red. The interference colors in the various quadrants increase or decrease from first-order red depending on whether the retardations in the accessory plate add to or subtract from the retardations in each quadrant. For positive minerals the N-W and S-E quadrants turn yellow while the N-E and S-W quadrants turn blue. The opposite occurs with a negative mineral.

order blue. If the mineral had been negative, the N-W quadrant would have increased to blue and the N-E quadrant would have decreased to yellow. Minerals with high birefringence will have many isochromatic rings, which may make the change to blue or yellow near the optic axis difficult to see. In such a case, the quartz wedge can be used to increase the changes in retardation.

Few grains in a thin section are likely to have the orientation shown in Fig. 2-6 (these will remain in extinction under orthoscopic light when the stage is rotated). If the indicatrix is only slightly tilted, the optic axis will still be present in the field of view. The center of the cross, however, will rotate around the axis of the microscope as the stage is rotated, but the isogyres will maintain their E-W or N-S orientation. Even if the center of the cross is outside the field of view (Fig. 2-7), its approximate location can be interpreted from the way the isogyres sweep across the field. The sign of the mineral can also be determined in these off-centered figures as long as quadrants are correctly identified. It should be emphasized that once the optic axis moves well out of the field it is difficult to distinguish a uniaxial mineral from a biaxial one with a small 2V.

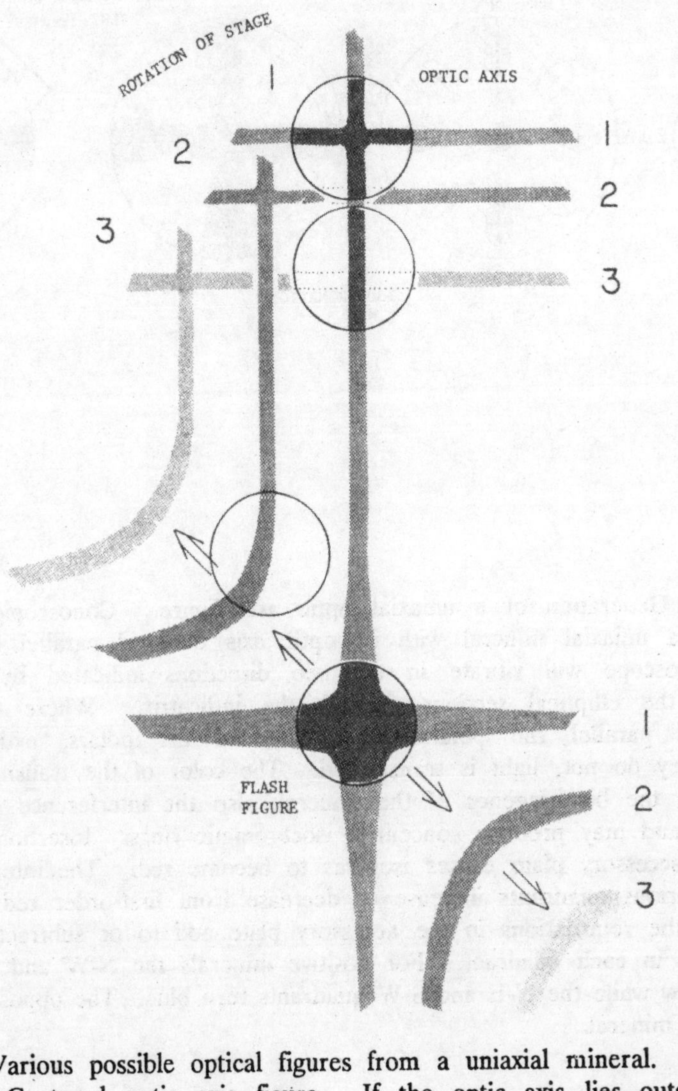

Figure 2-7 Various possible optical figures from a uniaxial mineral. From top to bottom: Centered optic axis figure. If the optic axis lies outside the field of view, a straight isogyre will sweep across the field as the stage is rotated, as shown in the successive positions 1, 2, and 3. If the optic axis is a long way out of the field of view, the isogyre will bend as the stage is rotated and the figure could be confused for a biaxial one. In the centered flash figure the diffuse cross breaks up and leaves the field of view after only a few degrees of rotation. The isogyres leave in those quadrants that contain the optic axes.

If the optic axis of a uniaxial mineral lies on the stage of the microscope, the interference figure obtained is known as a *flash figure*. Grains with this orientation exhibit the highest interference colors. When the optic axis parallels one of the polars, the isogyres form a centered diffuse cross. On rotating the stage, the isogyres separate quickly into two curved bands which leave the field after only a few degrees of rotation. As seen in Fig. 2-7, the isogyres leave the field of view in the quadrants that contain the optic axes. It is therefore possible to determine the sign of the mineral from a flash figure by noting whether the optic axis direction is the slow or fast vibration direction.

Off-centered figures that are nearer the optic axis than the flash figure position have straight isogyres that are oriented E-W or N-S. As the figure approaches the flash figure, however, the end of the isogyre nearest the flash figure shows distinct curvature and may be mistaken for a biaxial figure (Fig. 2-7). If there is any doubt about the identity of such a figure, it is best to search for a more favorably oriented grain.

A biaxial crystal having any one of the principal vibration directions of the indicatrix parallel to the axis of the microscope and the other two parallel to the polars will exhibit an interference figure that resembles the uniaxial optic axis figure, that is a cross. On rotating the stage, however, the cross breaks up into two isogyres. The resulting figure depends on the orientation of the indicatrix. The main types of interference figure and their relation to the indicatrix are shown in Fig. 2-8 for a crystal that has been rotated into the 45° position.

If the y vibration direction parallels the axis of the microscope, a *flash figure* is obtained. This resembles the uniaxial flash figure in that the isogyres leave the field completely following a stage rotation of only a few degrees (less for crystals with 2V near 90°). Flash figures are useful, for they identify those crystals that have the x and z vibration directions on the stage of the microscope. Measurements of extinction angles, for example, commonly require that grains have this orientation.

If the z or x vibration direction parallels the axis of the microscope, a *bisectrix* figure is obtained. The bisectrix can be acute or obtuse depending on the optic angle. When 2V = 90°, there is no distinction between the acute (BXA) and obtuse (BXO) figures. Figure 2-8 illustrates a positive mineral with a moderate 2V; when the z vibration direction parallels the axis of the microscope, a BXA is obtained; when x parallels the axis of the microscope, a BXO is obtained.

As the stage is rotated to the 45° position, the cross of a centered BXA opens slowly (not rapidly, as with a flash figure) as the isogyres separate into the quadrants containing the optic axes. The optic axes can be identified as the points of greatest curvature of the isogyres, and if the mineral is strongly birefringent, isochromatic lines will form a figure eight pattern with the optic axes at the centers of the two circles. As long as the optic angle remains less than about 55° (this value depends on the lens aperture and the R.I. of the mineral), the optic axes will remain in the field of view. If they just leave the field of view, 2V must be slightly greater than 55°. If, however, they leave the field of view well before the stage has been rotated 45°, 2V must be very large. In such cases, there may be uncertainty as to whether you are dealing with a BXA or BXO figure. Misidentification of the figure can lead to an erroneous sign determination. It is therefore preferable to seek an optic axis figure.

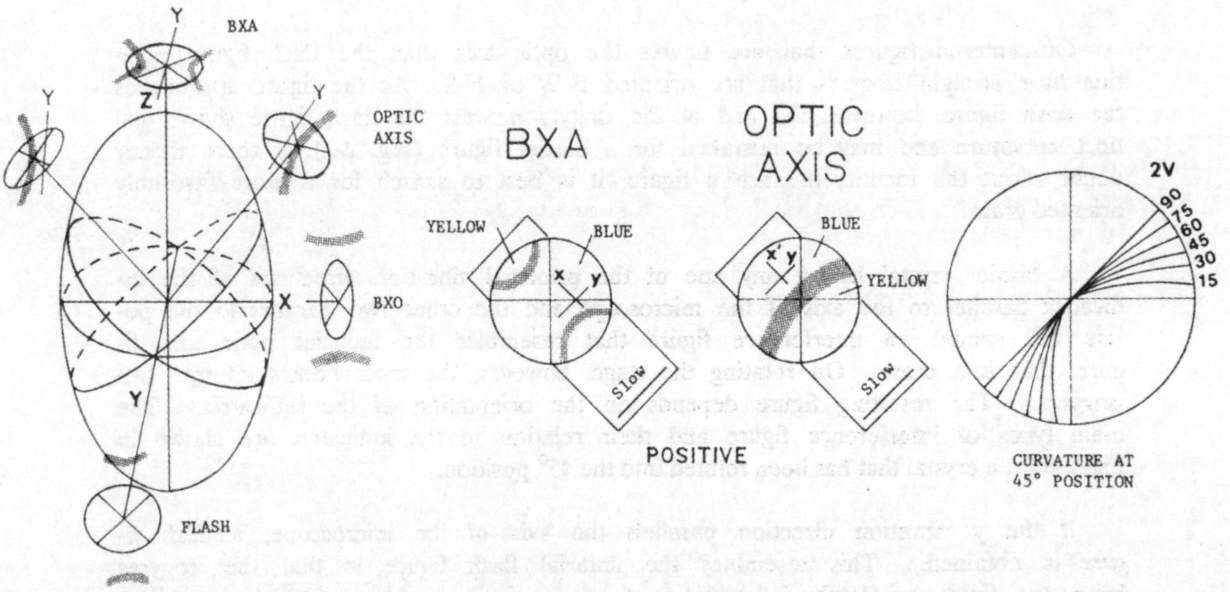

Figure 2-8 Main types of optical figures associated with a biaxial mineral. All are shown rotated to the 45° position. Sign determinations are easily made on the acute bisectrix (BXA) and optic axis figures, because the y vibration direction is tangential to the isogyre at its point of maximum curvature when the figure is in the 45° position. The identity of the other vibration direction (x or z) can be determined with an accessory plate. In this example, the mineral is positive (see text for explanation). The 2V can be estimated from the curvature of the isogyre in the 45° position.

With a bisectrix figure in the 45° position, the y vibration direction lies between the optic axes and is tangential to the isogyres at the optic axes (Fig. 2-8). Whether y is faster or slower than the other vibration direction can be determined with an accessory plate. In this example, the area between the isogyres would turn blue, and therefore y is the slow vibration direction. The other vibration direction must be x. By elimination, then, z must parallel the axis of the microscope and lie in the acute bisectrix. The mineral is therefore positive.

A centered optic axis figure consists of a single isogyre, which rotates around like a propeller as the stage is rotated. The isogyre may curve as the crystal is rotated into the 45° position. The amount of curvature depends on the optic angle (Fig. 2-8). If 2V = 90°, the isogyre is straight; if 2V = 0° (uniaxial), the isogyre curves through 90°. The curvature of the isogyre for various 2V's is given in Fig. 2-8. Because, in the 45° position the isogyre is convex towards the BXA, and the y vibration direction is tangential to the isogyre at the optic axis, the sign can be determined in the same way as for a BXA figure. If it cannot be determined which side of the isogyre is convex, the sign cannot be determined. This will occur only when 2V approaches 90°, in which case there is no significance to determining sign.

Many different off-centered figures can be obtained on biaxial minerals, some of which may be difficult to identify uniquely. If a figure is misidentified, erroneous observation of optical properties will result. It is prudent, therefore, to search for grains with other orientations if there is any doubt as to the identity of a figure.

The refractive indices of minerals change with the wavelength of light, as is evident from the splitting of white light into the visual spectrum on passing through a glass prism. When the refractive indices of a mineral are listed with no reference to wavelength, it is understood that the values are for the dominant wavelength of sodium light, which is 589 nm. The vibration directions of light also can vary with changes in wavelength. Any change in optical properties resulting from changes in the wavelength of light is known as *dispersion*.

In biaxial minerals, dispersion can lead to striking optical effects in interference figures. If the three refractive indices vary differently with wavelength, a different indicatrix exists for each wavelength. This, in turn, causes the optic angle to change with wavelength. If white light is used, the optic angle for wavelengths at the red end of the spectrum may be larger or smaller than that for wavelengths at the blue end. If, for example, the optic angle for blue light is greater than that for red light, the blue isogyre would occur farther out than the red one in a BXA figure rotated to the 45° position. But the blue isogyre marks those locations in the figure where blue light is in extinction. If no blue is transmitted, the remaining light in this region will appear red. Likewise, in the region of the red isogyre the transmitted light will appear blue. Thus, when the blue optic axis is greater than the red one, a red fringe will appear on the outside (concave) of the isogyres in a BXA figure in the 45° position, and a blue fringe will appear on the inside (convex) of the isogyres. The dispersion in such a mineral would be described as being less for red light than for blue, or $r < b$. More complex dispersions involving changes in the vibration directions occur in some minerals belonging to the monoclinic and triclinic systems. Such dispersion is not commonly used to identify minerals but in rare cases can be diagnostic. Readers are referred to texts on optical mineralogy for further information about dispersion.

Observation of Crystals under Reflected Light

A number of important minerals, including many of economic value, are opaque to visible light. Little can be learned about these minerals in thin sections from studies in transmitted light, except possibly their external morphology and alteration products, which may be transparent. Sulfides can often be distinguished from opaque oxides in ordinary thin sections by their yellowish color when illuminated obliquely from above. To properly identify opaque minerals, however, uncovered polished thin sections must be used with a reflecting microscope. Many petrologists use such sections routinely, because they are also necessary if analytical work is to be done with the electron microprobe. This instrument bombards the polished surface of a mineral with high-energy electrons (15-20 kv) which cause the elements present to emit x-rays. From a determination of the intensity of x-rays with wavelengths characteristic of the particular elements, a quantitative analysis of a small (<10 μm) spot is obtained.

If polished thin sections are used for study with a normal transmitted light microscope, it is important to remember that the resolution of high power objectives will be poor without a cover glass. If high resolution is required, a glass cover can be placed on the thin section temporarily, using immersion oil as the mounting medium. Alternatively, a reflecting light objective, which is designed to be used without a cover glass, can be used.

The reflecting microscope differs from the transmitted one in that the objective lens is used both to view and to illuminate the sample (Fig. 2-9). A beam of light, which may be plane-polarized by insertion of a polar, is reflected down through the objective onto the surface of the polished section by a prism or a coated cover glass placed in the optical path of the microscope above the objective lens. The image of the polished surface is then transmitted back through the objective to the ocular. The illuminating assembly has two diaphragms. The one nearest the microscope (field diaphragm) should be closed down until the edge of the iris appears at the margin of the field of view. Stopping down the other diaphragm (aperture diaphragm) will bring out greater contrasts in relief, reflectivity, and color of minerals. To observe minerals under crossed polars, the analyzer is inserted in the same way as with transmitted light.

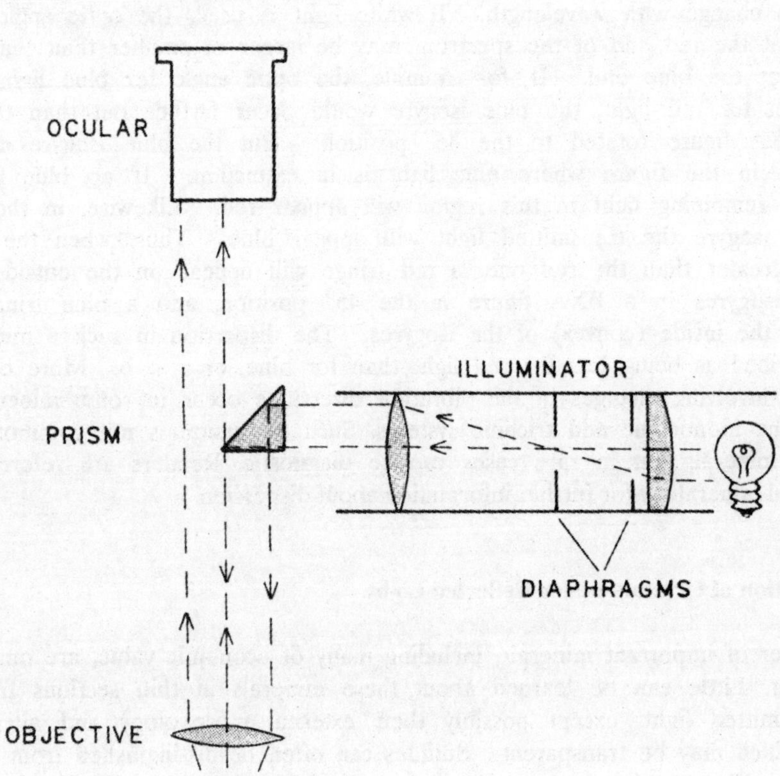

Figure 2-9 Basic arrangement of the optics in a reflecting microscope.

The amount of light reflected from the polished surface of a mineral is known as its *reflectivity*. Obviously opaque minerals are more reflective than transparent ones, but even among the opaque minerals there is considerable variability. The value of the reflectivity can be measured accurately with photometers, but for routine petrography, visual estimates of the brightness of minerals are adequate.

If plane-polarized white light is used, the intensity and color of the reflected light from anisotropic minerals may change with orientation. This gives rise to *bireflectance*, which is equivalent to absorption and pleochroism in transmitted light. The colors encountered with reflected light, however, are far less vivid than those with transmitted light. Indeed, when first encountered, most opaque minerals appear almost the same color in reflected light--some shade between white, gray, and cream. With experience, however, subtle differences become far more easily recognized.

Under crossed-polars, isotropic opaque minerals can be distinguished from anisotropic ones. Here, again, the interference colors are far less vivid than those obtained with transmitted light. Nonetheless, anisotropic minerals under reflected light will brighten and darken four times while the stage is rotated through 360 degrees, unless, of course, the optic axis of the mineral is oriented parallel to the microscope axis. Commonly, the polars are slightly uncrossed (few degrees), allowing some light to pass through the microscope and illuminate the field. In this position, however, the polars do not cause a complete extinction, only a darkening of the field. Nor will they cause the extinction of isotropic minerals; but isotropic minerals will remain an unchanging dark color as the stage is rotated.

Although refractive indices of opaque minerals are not measured, it is possible to observe differences in hardness between minerals simply from the relief on the polished surface. Accurate hardnesses can be measured with microhardness testers. But for routine work, relative hardnesses are adequate. By increasing the distance between the sample and the objective, a bright line of light, similar in appearance to the Becke line, will move towards the softer grain. Hardnesses of opaque minerals are commonly given in terms of the Talmage scale, rather than the Mohs scale. The Talmage scale has seven divisions, designated by letters, with A being the softest and G the hardest.

Preparation of Thin Sections

A petrographic thin section is a 30 μm-thick slice of rock mounted on a glass microscope slide and either covered with a cover glass or left uncovered but polished. Although sections can be thicker or thinner than this, listed interference colors for minerals in standard tables and in this manual are for 30 μm-thick grains. This thickness (or thinness, if you are struggling to make your first section) is easily attainable with equipment that is in proper alignment. In sections that are thinner than standard, the decreased contrast in interference colors between different minerals makes identification more difficult. And in sections that are thicker than normal, grain boundaries that are not vertical appear as fuzzy zones instead of distinct lines. Indeed, with fine-grained rocks, even a 30 μm-thick section may not produce a distinct image of the grains. In extremely fine-grained rocks, a distinct image may be obtainable only in reflected light from the polished surface.

To prepare a thin section, a slab of rock approximately 22 mm x 40 mm x 10 mm thick is cut with a diamond saw. The surface that is to be glued to the glass slide is then ground successively with 240, 600, and 1000 grit carborundum powder. The purpose of this is to obtain a perfectly flat surface. This stage of the process is one of the most critical, and yet it is often rushed and not done properly. If the surface that is glued to the slide is curved, there is nothing that can be done later to salvage the section. Even if the upper surface of the slide is perfectly flat, the beveled lower surface at best will cause variations in the thickness of the section (hence of interference color), and at worst will result in the complete grinding away of parts of the section. Grinding with 1000 grit powder can be omitted, but it does produce cleaner-looking sections and does help prevent plucking of easily cleaved minerals (phenocrysts of amphibole, for example). If extremely fine features are being sought in a thin section, the lower surface of the rock chip can even be polished before it is glued to the slide. This would be done only if the upper surface is also to be polished. It should be noted, however, that in doubly-polished sections, the apparent roughness of high relief minerals is not visible; feldspar and olivine, for example, would appear to have the same low relief.

The rock chip is glued to a petrographic glass slide (typically measuring 27 x 46 mm) with epoxy. Traditionally Canada balsam was used for this purpose, but now this natural glue is used only for attaching cover glasses; epoxies are simply much stronger. Nonetheless, it is desirable to have an epoxy whose refractive index is similar to that of Canada balsam (R.I. = 1.537), because the relief described for minerals in the literature is given with respect to Canada balsam. A number of epoxies are commonly used, for example, Petropoxy 154 of Palouse Petro Products, Buehler 20-8130, and Hillquist A-B and C-D. Only Petropoxy has a refractive index of 1.54; the others are all between 1.57 and 1.58. This is particularly important to keep in mind when identifying minerals on the basis of relief.

Epoxies cure more rapidly when heated, but most will cure at low temperature, if given time. Petropoxy, however, must be heated in order to harden. Rock chips are heated before the epoxy is applied, not only to raise the rock to the recommended curing temperature, but to drive off moisture from the sample. Water given off after the epoxy has been applied forms large bubbles that may cause the rock to detach from the slide during the subsequent cutting and grinding steps. Some rocks contain minerals that continuously give off water as they are heated, for example, zeolites and gypsum. Rocks containing such minerals may have to be mounted with cold setting epoxy if gas bubbles keep forming on heating. One particular advantage in using heated epoxy is that it has substantially lower viscosity, which allows the excess epoxy between the sample and the slide to be squeezed out more easily by placing a weight on the rock or clipping the slide and rock together with a spring-backed paper clip. Variable thicknesses of glue on different sections slow up the process of section making, because each section has to be checked individually many more times than are ones that all have a constant thickness of glue. A more serious problem, however, arises with those sections in which variations in the thickness of glue cause the thin section to be wedge shaped. Such sections may require extensive final grinding by hand in order to compensate for the rock chip not being mounted parallel to the glass slide.

Once the epoxy has hardened and the rock chip cooled to room temperature, the mounted specimen is placed in a cut-off saw and sliced so that an approximately 250μm-thick section is left attached to the glass slide. The initial slice can be thinner than this if the equipment is in good alignment; but the next, and final, step goes so quickly that there seems no justifiable reason for pushing one's luck by cutting extra thin sections at this stage. Commonly, when the initial section is cut too thin, vibrations from the saw blade may loosen the sample from the slide or fracture unnecessarily coarse cleavable mineral grains which will later pluck out.

Next the section is ground, either by hand, or preferably on a grinding machine. When a thickness of 30 μm is approached, the section is removed from the grinder and the interference colors examined under crossed polars in the microscope. Interference colors provide a very simple but extremely accurate means of judging section thickness. This does, however, require that the section maker be able to identify the minerals, know what interference colors they should exhibit in a 30 μm-thick section, and be aware of the effect of grain orientation on interference colors. For the novice, a micrometer is simpler to use. The final grinding steps are repeated until the correct thickness is obtained. If a polished thin section is to be prepared, the grinding is stopped when the section is ~35 μm thick. The following step with polishing powder removes the final 5 μm. If a normal thin section is to be prepared, a cover glass is mounted with Canada balsam, which is then cured at approximately 90°C for approximately half an hour. Petropoxy can also be used for this purpose, because it is extremely fluid when hot, and it hardens in just 3 minutes at 125°C.

General Hints on Doing Petrography

Perhaps through an earnest desire to apply everything learned in optical mineralogy, the novice petrographer tends to start viewing a thin section at too high a magnification and with too many accessories inserted into the microscope. The experienced petrographer, by contrast, uses the lowest possible magnification commensurate with the grain size of the rock and examines the rock mainly under plane polarized light or possibly crossed polars. Of course, when first learning petrography a high power objective may be necessary to determine properties that will confirm the identity of a mineral. With experience, however, most of the common rock-forming minerals can be distinguished simply under plane-polarized light.

Start your inspection of the thin section under a low-power objective and with plane-polarized light. You should distinguish felsic minerals (low relief--commonly negative, and mostly colorless) from ferromagnesian minerals (high relief, mostly colored). In addition, features such as pleochroism, grain shape, textural relations between grains, and cleavage can also be noted. Only then should the polars be crossed. Do not lose track of the identity of the felsic and ferromagnesian minerals while examining the section under crossed polars. For example, orthopyroxene and plagioclase, which commonly occur together in a type of gabbro known as norite, have very similar interference colors and may appear similar under crossed polars. The difference in relief of the two minerals, however, makes them quite distinct under plane-polarized light. Finally switch to high magnification and conoscopic light if it is necessary to obtain an interference figure.

The light source on the microscope should not be turned up to its brightest setting for the initial petrographic survey, nor should the upper condenser be inserted. These will decrease the contrast in color and relief between minerals.

It is not necessary to determine all of the optical properties of a mineral to confirm its identity--this would make a petrographic study of a thin section particularly time consuming. Part of the task of learning petrography is to know which properties need to be measured for a particular mineral. The descriptions in the following chapter emphasize the most diagnostic properties of minerals. Although you will tend to overdetermine minerals at first, experience will give you the necessary confidence to reduce your observations to the minimum.

Experienced petrographers keep a mental tally of each mineral they identify in a rock. This list provides them with guidelines when they attempt to identify the next unknown mineral in the specimen. Certain minerals are commonly associated, whereas others never occur together. For example, in tholeiitic rocks, augite is extremely common with orthopyroxene. On the other hand, a rock containing nepheline could not contain quartz--the two minerals would react and form albite. In metamorphic rocks, the number of minerals tends to be limited by the number of components present (generally the two are equal); this may help you decide whether to look for additional minerals, or perhaps you have found too many, and two of your identified phases may be differently oriented grains of the same mineral. This approach to thin section study, however, is only possible after you have acquired the necessary petrologic knowledge.

A suggested format for recording petrographic observations is given on the last page of this manual. This form fits conveniently on 5 x 8 inch index cards. It can be photocopied or modified to suit your specific needs.

3 Rock-Forming Minerals and Their Optical Properties

This chapter contains a tabulation of the common minerals and their formulae, followed by a listing of their most important distinguishing optical properties. Most of the minerals are listed alphabetically for ease of referencing. Exceptions occur where minerals belong to well-recognized groups. Thus, all Al_2SiO_5 polymorphs are listed together under Al_2SiO_5. Other minerals listed under a common heading include amphiboles, carbonates, epidotes, feldspars, garnets, micas (+ pyrophyllite), olivines, opaques, pyroxenes, silica polymorphs (under quartz), spinels, and zeolites (+ analcite). For example, the optical properties of the amphibole grunerite are listed with other amphiboles under **AMPHIBOLE** rather than alphabetically under **G**. If a mineral cannot be found under the alphabetical listing and you are uncertain to which mineral group it belongs, the mineral can be found in the index, if it is included in the manual. A blank page is provided at the end of this chapter for the addition of optical properties of less common minerals that you may wish to add to the manual.

The optical properties of each mineral are given in a standard format which is self explanatory. Illustrations of the relation between optics and crystal morphology are included for those minerals where this information is useful. If the mineral is illustrated in the thin section sketches (and color slides), the number of the illustration is given in parentheses beneath the mineral name, with IM# referring to the sketches of igneous minerals and MM# to those of metamorphic minerals.

In addition to the listings of optical properties of each mineral, a compilation of the most important properties of all the minerals is presented in a single table on the inside front cover of the manual. For most routine inspections of thin sections, this table provides most of the information necessary to identify the common minerals. It is worth spending a few minutes familiarizing yourself with its organization.

The table is divided down the center by a chart of interference colors. To the left are listed minerals which are colored under plane-polarized light, and to the right are minerals which are colorless. Some minerals appear on both sides if they are only faintly colored, or they are colorless for some composition but colored for others. Both groups of minerals are listed in order of increasing birefringence, with isotropic ones given first. The interference colors in the central strip are those found in normal 30 μm-thick thin sections for the particular birefringence indicated.

The refractive index range for each mineral is indicated by a bar with the actual values being indicated across the top of the table. The relief relative to the refractive index of Canada balsam (1.537), is indicated by the density of stippling. The refractive index bars in the left side of the table are colored to indicate the typical appearance of these minerals in plane polarized light; pleochroic minerals have two colors.

The table also gives the optical character of the mineral--isotropic (I), uniaxial (U), and biaxial (B)--optic sign, optic angle, and prominent cleavage planes and the angles between them.

IMPORTANT MINERALS AND THEIR FORMULAE

The following minerals are those most commonly encountered in rocks. They are divided into three groups, the first two including the rock-forming minerals, and the third, the accessory minerals. The rock-forming minerals are the major building components of a rock, and they determine its composition and name. Minerals in the first group have low refractive indices (RI $\leq$ Canada balsam), and thus have low relief in thin section, and are colorless; they include quartz, feldspars, feldspathoids, and the one ferromagnesian mineral cordierite. Minerals in the second group have high refractive indices (RI >> Canada balsam) and thus have high relief in thin section, and most are colored; this includes the ferromagnesian minerals. The accessory minerals are rarely present in more than trace amounts, but their presence or absence can be critical to the interpretation of the genesis of a rock. Most accessory minerals are formed from elements that do not readily enter the structures of the major minerals.

I. ROCK-FORMING MINERALS WITH LOW REFRACTIVE INDEX (Colorless)

	Name	**Formula**
	Quartz	
	Tridymite	SiO_2
	Cristobalite	
	Sanidine	
	Orthoclase	$(K,Na)AlSi_3O_8$
FELDSPAR	Microcline	
	Albite	$NaAlSi_3O_8$
	Anorthite	$CaAl_2Si_2O_8$
	Nepheline	$(Na,K)AlSiO_4$
	Kalsilite	$(K,Na)AlSiO_4$
FELDSPATHOID	Leucite	$KAlSi_2O_6$
	Sodalite	$Na_8Al_6Si_6O_{24}Cl_2$
	Analcite	$NaAlSi_2O_6\ H_2O$
	Scapolite	$(Na,Ca,K)_4Al_3(Al,Si)_3Si_6O_{24}(Cl,CO_3,SO_4,OH)$
	Cordierite	$(Mg,Fe)_2Al_4Si_5O_{18}$

II. ROCK-FORMING MINERALS WITH HIGH REFRACTIVE INDEX (Mostly colored)

	Name	Formula
OLIVINE	Forsterite	Mg_2SiO_4
	Fayalite	Fe_2SiO_4
	Monticellite	$CaMgSiO_4$
ORTHOPYROXENE	Enstatite	$Mg_2Si_2O_6$
	Ferrosilite	$Fe_2Si_2O_6$
CLINOPYROXENE	Diopside	$CaMgSi_2O_6$
	Hedenbergite	$CaFeSi_2O_6$
	Augite	$(Ca,Mg,Fe,Al)_2(Si,Al)_2O_6$
	Pigeonite	$(Mg,Fe,Ca)(Mg,Fe)Si_2O_6$
	Aegerine (Acmite)	$NaFe^{+3}Si_2O_6$
	Jadeite	$NaAlSi_2O_6$
	Wollastonite	$CaSiO_3$
AMPHIBOLE	Anthophylite	$(Mg,Fe)_7Si_8O_{22}(OH,F)_2$
	Gedrite	$(Mg,Fe)_5Al_2(Al_2Si_6)O_{22}(OH,F)_2$
	Cummingtonite	$(Mg,Fe)_7Si_8O_{22}(OH,F)_2$
	Tremolite-Actinolite	$Ca_2(Mg,Fe)_5Si_8O_{22}(OH,F)_2$
	Hornblende	$Ca_2(Mg,Fe,Al)_5(Si,Al)_8O_{22}(OH,F)_2$
	Riebeckite	$Na_2Fe_3{}^{+2}Fe_2{}^{+3}Si_8O_{22}(OH,F)_2$
	Glaucophane	$Na_2Mg_3Al_2Si_8O_{22}(OH,F)_2$

II. Continued	Name	Formula
Micas	Biotite	$K(Mg,Fe)_3(AlSi_3O_{10})(OH,F)_2$
	Muscovite	$KAl_2(AlSi_3O_{10})(OH,F)_2$
	Paragonite	$NaAl_2(AlSi_3O_{10})(OH,F)_2$
	Pyrophyllite	$Al_2Si_4O_{10}(OH)_2$
	Talc	$Mg_3Si_4O_{10}(OH)_2$
	Chlorite	$(Mg,Al,Fe)_6(Al,Si)_4O_{10}(OH)_8$
	Serpentine	$Mg_6Si_4O_{10}(OH)_8$
Garnet	Pyrope	$Mg_3Al_2Si_3O_{12}$
	Almandine	$Fe_3Al_2Si_3O_{12}$
	Spessartine	$Mn_3Al_2Si_3O_{12}$
	Grossular	$Ca_3Al_2Si_3O_{12}$
	Andradite	$Ca_3(Fe^{+3},Ti)_2Si_3O_{12}$
	Vesuvianite(Idocrase)	$Ca_{19}(Mg,Fe,Al)_{13}Si_{18}(O,OH,F)_{76}$
	Andalusite Kyanite Sillimanite	Al_2SiO_5
	Mullite	$3Al_2O_3.2SiO_2$
	Staurolite	$Fe_2Al_9Si_{3.75}O_{22}(OH)_2$
	Chloritoid	$(Fe^{+2},Mg,Mn)_2(Al,Fe^{+3})Al_3O_2(SiO_4)_2(OH)_4$
	Epidote	$Ca_2Fe^{+3}Al_2O(Si_2O_7)(SiO_4)(OH)$
	Clinozoisite	$Ca_2AlAl_2O(Si_2O_7)(SiO_4)(OH)$
	Lawsonite	$CaAl_2(OH)_2Si_2O_7H_2O$
Melilite	Gehlenite	$Ca_2Al_2SiO_7$
	Akermanite	$Ca_2MgSi_2O_7$
	Soda melilite	$NaCaAlSi_2O_7$
	Calcite	$CaCO_3$
	Dolomite	$CaMg(CO_3)_2$

III. ACCESSORY MINERALS

	Name	Formula
	Apatite	$Ca_5(PO_4)_3(OH,F,Cl)$
	Zircon	$ZrSiO_4$
	Sphene	$CaTiSiO_5$
	Perovskite	$CaTiO_3$
	Tourmaline	$Na(Mg,Fe,Al)_3Al_6Si_6O_{18}(BO_3)_3(OH,F)_4$
	Corundum	Al_2O_3
	Rutile	TiO_2
	Hematite	Fe_2O_3
	Ilmenite	$FeTiO_3$
Spinel	Ulvospinel	Fe_2TiO_4
Spinel	Magnetite	Fe_3O_4
Spinel	Chromite	$FeCr_2O_4$
Spinel	Spinel	$MgAl_2O_4$
Spinel	Hercynite	$FeAl_2O_4$
	Fluorite	CaF_2
	Pyrite	FeS_2
	Pyrrhotite	Fe_7S_8 - FeS
	Chalcopyrite	$CuFeS_2$
	Sphalerite	ZnS
	Anhydrite	$CaSO_4$
	Gypsum	$CaSO_4.2H_2O$
	Barite	$BaSO_4$
	Beryl	$Be_3Al_2[Si_6O_{18}]$

ANDALUSITE Al_2SiO_5

Biaxial - Orthorhombic

(MM 9-12)

Refractive Index Relief	Birefringence Interf. Color	2V	Extinction Elongation
x 1.63-1.65	0.01		parallel
y 1.63-1.65		73-86	length fast
z 1.64-1.66	1st order white		

Color Pleochroism	Cleavage	Cleavage Angle	Hardness
colorless			
x	{110} good	{110}^{1$\bar{1}$0} 89	7
y			
z			

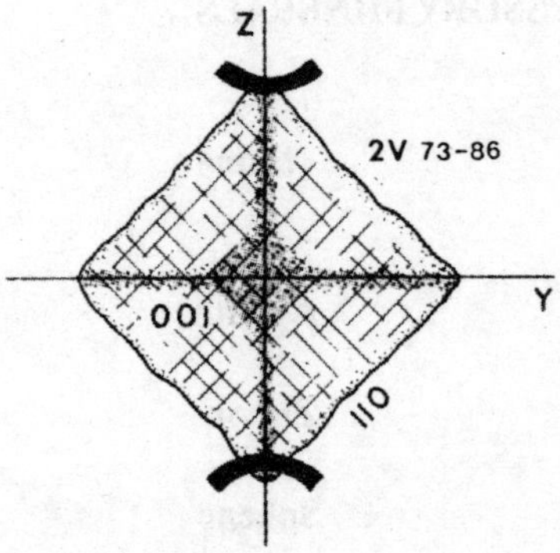

Morphology: Coarse columnar grains with almost square cross sections. Crystals may contain many inclusions, those of carbonaceous material commonly being concentrated into the core or along planes that represent the intersection of the prism faces as the crystal grew. Chiastolite is a variety in which the inclusions form a prominent cross.

Composition: Andalusite varies little in composition. Some varieties contain small amounts of ferric iron and Mn, which makes them faintly pleochroic in shades of pink and green.

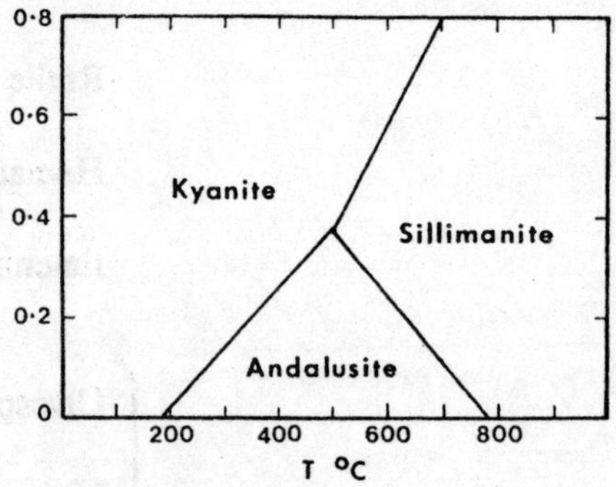

Phase relations among the Al_2SiO_5 polymorphs (Holdaway,1971).

Distinguishing Properties: Low birefringence and large 2V distinguish andalusite from sillimanite. Kyanite has inclined extinction and higher relief. Pleochroic varieties of andalusite resemble hypersthene, but hypersthene is length slow. Apatite is uniaxial. Under crossed polars, large inclusion-filled porphyroblasts of andalusite resemble those of cordierite, but under plane-polarized light andalusite has higher relief. Commonly altered to muscovite.

Occurrence: Forms in metamorphosed pelitic rocks at low to medium grades of regional metamorphism. Its stability is limited to pressures of less than 0.375 GPa (3.75 kbar); above this it changes to the polymorph kyanite, and at high temperatures, to the polymorph sillimanite, as shown in the adjoining figure. Andalusite is a common contact metamorphic mineral in pelitic rocks around high-level plutons.

KYANITE Al_2SiO_5

Biaxial - Triclinic

(MM 5-10)

Refractive Index Relief	Birefringence Interf. Color	2V	Extinction Elongation
x 1.71	0.015		z^c 30
y 1.72		82	length slow
z 1.73	1st order red		

Color Pleochroism	Cleavage	Cleavage Angle	Hardness
colorless			
x	{100} perfect	>	5 and 7
y	{010} good	> all 85-90	
z	{001} parting	>	

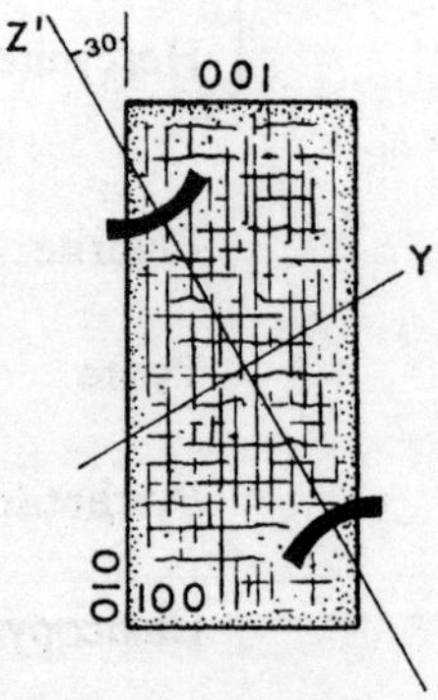

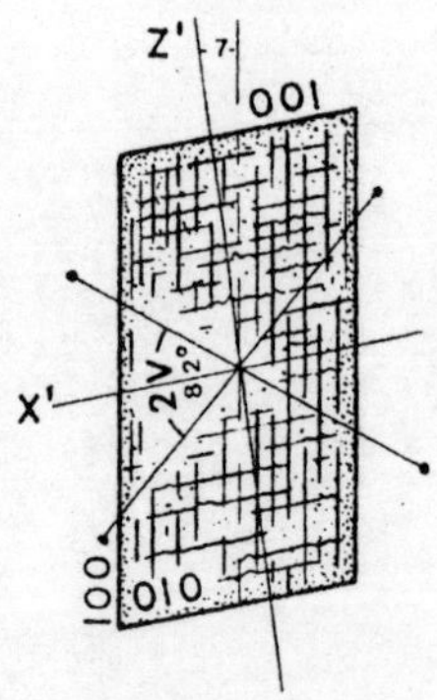

Morphology: Bladed crystals which apppear broad when viewed on (100)--these sections have an extinction at 30° to the length--and narrow when viewed on (010)--these sections have extinction almost parallel to the length.

Composition: Kyanite shows almost no variation in composition.

Distinguishing Properties: Although colorless, the high relief makes kyanite appear quite dark in thin section, specially along grain boundaries and cleavages. Cleavages and cross parting are prominent. The variable extinction angle (0-30) of the slow vibration direction on the length of crystals is characteristic of kyanite.

Occurrence: Strictly a metamorphic mineral formed during regional metamorphism of pelitic rocks. It is the high-pressure aluminum silicate polymorph and is stable at intermediate grades of metamorphism. Depending on the nature of the geothermal gradient during metamorphism, andalusite may be present at lower grades, whereas the high-temperature polymorph, sillimanite, is normally stable at higher grades.

SILLIMANITE Al_2SiO_5

Biaxial + Orthorhombic

(MM 1-4)

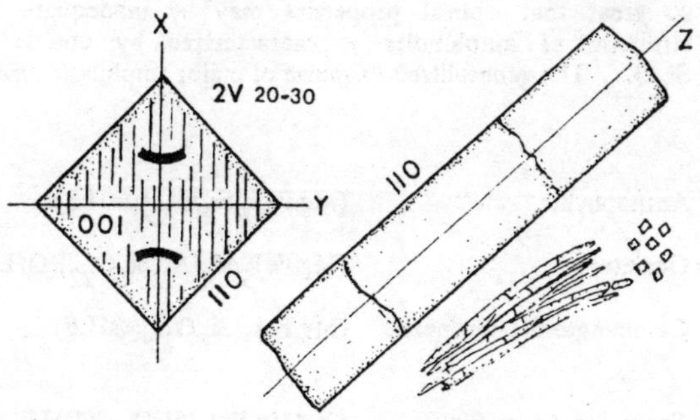

Refractive Index Relief	Birefringence Interf. Color	2V	Extinction Elongation
x 1.65-1.66	0.02		parallel
y 1.66		20-30	length slow
z 1.67-1.68	2nd order blue		

Color Pleochroism	Cleavage	Cleavage Angle	Hardness
colorless			
x	{010} good		7
y			
z			

Morphology: Prisms or fibrous aggregates (fibrolite). Prisms have square cross sections with diagonal {010} cleavage.

Composition: As with the other Al_2SiO_5 polymorphs, sillimanite differs little from the formula.

Distinguishing Properties: Slender prisms with widely-spaced cross fractures, parallel extinction, length slow, moderate birefringence, and strong dispersion (r>b) distinguish sillimanite. Apatite needles have much lower birefringence, are length fast, and have hexagonal cross sections. Anthophyllite has a large 2V (70-90), and in cross sections has the typical two cleavage directions of amphibole. Fibrous sillimanite may be more difficult to distinguish from fibrous anthophyllite; cross sections of prisms are the most diagnostic property in this case. Sillimanite is difficult to distinguish from mullite. Mullite is, however, a relatively rare mineral restricted to high-level, high-temperature contact metamorphosed pelites (specially as xenoliths in basic igneous rocks); it commonly has a faint lilac pleochroism.

Occurrence: Occurs in high-grade regional and contact metamorphic rocks. At lower temperatures kyanite and andalusite are stable, but inversion of sillimanite to these polymorphs on cooling is uncommon because of sluggish reactions.

MULLITE $3Al_2O_3.2SiO_2$

Biaxial + Orthorhombic

(MM 13, 14)

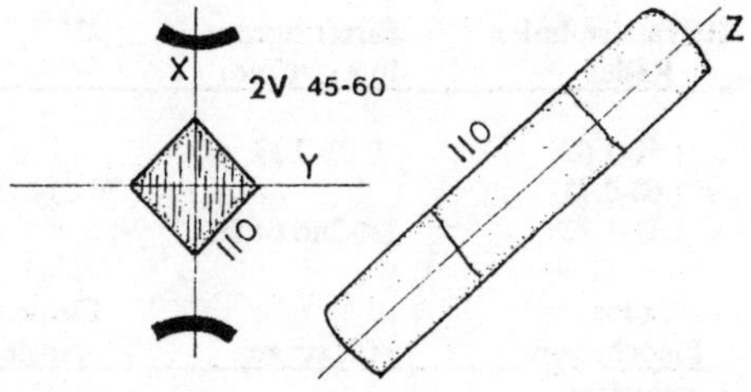

Refractive Index Relief	Birefringence Interf. Color	2V	Extinction Elongation
x 1.64-1.67	0.01-0.03		parallel
y 1.64-1.67		45-60	length slow
z 1.65-1.69	2nd order blue		

Color Pleochroism	Cleavage	Cleavage Angle	Hardness
colorless			
x colorless	{010} good		7
y colorless			
z pale lilac			

Morphology: Slender prisms with square cross sections and diagonal {010} cleavage.

Composition: Mullite is not a polymorph of Al_2SiO_5, for it is more aluminous and has a composition near $3Al_2O_3.2SiO_2$. It can also contain some ferric iron and Ti.

Distinguishing Properties: Properties are very similar to those of sillimanite, but mullite can contain more iron which makes it pleochroic. Mullite is rare, and its restricted mode of occurrence may help distinguish it from sillimanite.

Occurrence: Restricted to high-temperature low-pressure contact metamorphic pelites. It typically occurs in xenoliths in mafic igneous rocks.

Amphiboles are of three general types, Ca-poor, Ca-rich, and Na-rich. The Ca- and Na-rich varieties exhibit extensive solid solution but only limited solubility with the Ca-poor varieties. Compositional variability among amphiboles is so great that optical properties may be inadequate to do more than assign the amphibole to a general group. The structure of amphiboles is characterized by double chains of SiO_4 tetrahedra; the basic amphibole formula contains Si_8O_{22}. The generalized formulae of major amphibole types and some important optical properties are as follows:

		Symmetry	Sgn	2V	z^c	Color
Anthophylite	$(Mg,Fe)_7[Si_8O_{22}](OH,F)_2$	Orthorhombic	-+	70-110	0	colorless
Gedrite	$(Mg,Fe)_5AL_2[AL_2Si_6O_{22}](OH,F)_2$	Orthorhombic	+	80-90	0	colorless
Cummingtonite-Grunerite	$(Mg,Fe)_7[Si_8O_{22}](OH,F)_2$	Monoclinic	+-	65-90	21-10	colorless
Tremolite-Actinolite	$Ca_2(Mg,Fe)_5[Si_8O_{22}](OH,F)_2$	Monoclinic	-	86-65	20-12	colorless-green
Hornblende	$Ca_2(Mg,Fe,Al)_5[(Si,Al)_8O_{22}](OH,F)_2$	Monoclinic	-+	50-90	15-30	brown and green
Riebeckite	$Na_2Fe_3Fe_2[Si_8O_{22}](OH,F)_2$	Monoclinic	-	50-90	5	blue
Glaucophane	$Na_2Mg_3Al_2[Si_8O_{22}](OH,F)_2$	Monoclinic	-	0-50	5	pale blue

In addition to substitutions of Fe for Mg, Na for Ca, and Al for Si in the above formulae, K and Mn can substitute for Ca, and Fe''', Al, Ti, Mn, and Cr for Mg (specially in hornblende); many of these substitutions are coupled with each other and with Al substitution for Si in order to maintain charge balance. Substitution of some of these elements can be substantial, as for example in kaersutite (10 wt% TiO_2) in which some Ti may even substitute for Si. In some hornblende OH and F can be replaced by Cl and in basaltic hornblende and kaersutite by O. This latter substitution also involves the replacement of Fe'' by Fe''', and consequently these amphiboles are deep reddish brown.

Amphiboles are one of the most common groups of minerals, occurring widely in metamorphic and plutonic igneous rocks. They are major constituents of metamorphosed mafic igneous rocks and impure dolomitic limestones. Their composition can serve as a useful indicator of metamorphic grade. For example, in the lower part of the amphibolite facies tremolite-actinolite is stable, but with increasing grade more aluminum enters the amphibole and it changes to hornblende. Glaucophane, on the other hand, occurs in rocks metamorphosed at high pressures and low temperatures and imparts the blue color after which the blueschist facies is named. Hornblende is particularly common in diorites, granodiorites, granites, syenites, and monzonite. It is also common in alkaline mafic and even ultramafic igneous rocks where it may contain considerable Ti.

ANTHOPHYLLITE $(Mg,Fe)_7[Si_8O_{22}](OH,F)_2$

Biaxial - + Orthorhombic

GEDRITE $(Mg,Fe)_5Al_2[(Al,Si)_2Si_6O_{22}](OH,F)_2$

Refractive Index Relief	Birefringence Interf. Color	2V	Extinction Elongation
x 1.60-1.69	0.01-0.03		parallel c axis
y 1.60-1.71		70-110	length slow
z 1.61-1.72	1st-2nd order		

Color Pleochroism	Cleavage	Cleavage Angle	Hardness
colorless			
x -	{210} perfect	{210}^{2̄10} 54	6
y -			
z -			

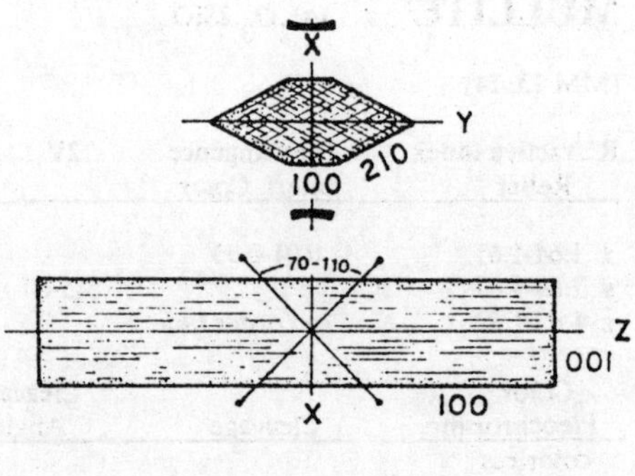

Morphology: Characteristically forms long prisms and may be fibrous.

Composition: Anthophyllite is aluminum-poor and gedrite is aluminum-rich. Compositions more iron-rich than 40 mole % Fe/(Fe+Mg) are monoclinic and belong to the cummingtonite-grunerite series.

Distinguishing Properties: Characterized by amphibole cleavage (54°) in sections cut across the c axis, by slender prismatic to fibrous habit, and parallel extinction in prismatic sections. Tremolite, cummingtonite, and grunerite have higher birefringence and have inclined extinction to the c axis. Anthophylite is distinguished from sillimanite in basal sections by the amphibole cleavage, and by its much larger 2V (sillimanite 2V=20-30). Distinction between anthophyllite and gedrite may require chemical analysis. Most anthopyllite is optically negative except for Fe-rich varieties which have high refractive index. Gedrite has a large 2V and is positive.

Occurrence: Anthophyllite is common in metamorphosed ultramafic rocks. It also occurs in Mg-rich metasomatic rocks associated with cordierite. This assemblage is found in some sulfide ore bodies. Introduction of sulfur into a rock can remove iron from the silicate minerals thus enriching them in Mg. Gedrite is common in meta-basalt.

CUMMINGTONITE-GRUNERITE $(Mg,Fe)_7[Si_8O_{22}](OH,F)_2$

Biaxial +- Monoclinic

(MM 48)

Refractive Index Relief	Birefringence Interf. Color	2V	Extinction Elongation
x 1.63-1.70	0.02-0.04		z^c 21-10
y 1.64-1.71		65-90	length slow
z 1.65-1.73	2nd order		

Color Pleochroism	Cleavage	Cleavage Angle	Hardness
colorless			
x -	{110} good	{110}^{1$\bar{1}$0} 55	6
y -			
z -			

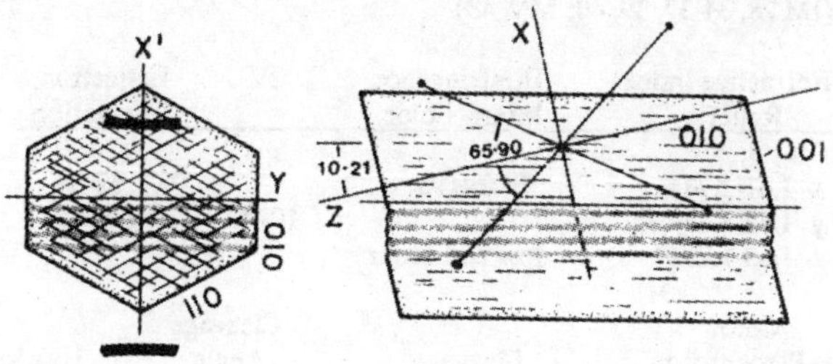

Morphology: Prismatic to fibrous. Prisms may be curved.

Composition: The mole % Fe/(Fe+Mg) in cummingtonite ranges from 30 to 70 and in grunerite, from 70 to 100. The 2V increases from +65 at the Mg-rich end to 90 at Fe/(Fe+Mg) of 0.75; from 0.75 the sign is negative, and the 2V decreases from 90 to 80 at the Fe end-member, grunerite.

Distinguishing Properties: Cummingtonite and grunerite are distinguished from anthophyllite by inclined extinction. Cummingtonite and grunerite commonly have fine multiple lamellar twins on {100}; twins are less common in tremolite and most are the simple type. The positive sign of cummingtonite also distinguishes it from tremolite. Cummingtonite is commonly intergrown with green hornblende and both amphiboles may contain exsolution lamellae of the other. The lamellae are orientated approximately parallel to {001}, and if twins are present, they may produce a herring-bone pattern.

Occurrence: Cummingtonite and grunerite occur principally in metamorphic rocks. Cummingtonite is commonly intergrown with green hornblende in metamorphosed mafic igneous rocks. Grunerite is common in Precambrian metamorphosed iron formations.

TREMOLITE-ACTINOLITE $Ca_2(Mg,Fe)_5[Si_8O_{22}](OH,F)_2$

Biaxial - Monoclinic

(MM 33, 34, 46, 47)

Refractive Index Relief	Birefringence Interf. Color	2V	Extinction Elongation
x 1.60-1.69	0.03-0.02		z^c 20-10
y 1.61-1.70		85-65	length slow
z 1.62-1.70	2nd order red		

Color Pleochroism	Cleavage	Cleavage Angle	Hardness
colorless to green			
x colorless-yellow	{110} good	{110}^{1$\bar{1}$0} 56	6
y colorless-green			
z colorless-green			

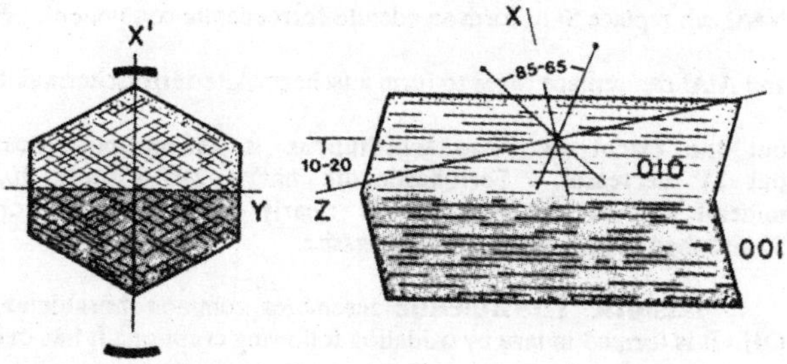

Morphology: Prismatic to fibrous aggregates.

Composition: With increasing substitution of Fe for Mg, the color changes from colorless to green, and the extinction angle and 2V decrease.

Distinguishing Properties: Tremolite is distinguished from cummingtonite by its negative optic sign. Also, multiple twins are not as common in tremolite as in cummingtonite. Actinolite is difficult to distinguish from hornblende and optical properties grade from the non aluminous actinolite to the aluminous hornblende. In general, extinction angles become larger with increasing aluminum, reaching 30 degrees in common hornblende.

Occurrence: Tremolite-actinolite is restricted to low grade metamorphic rocks. It forms in metamorphosed limestones and metamorphosed ultramafic rocks.

HORNBLENDE $(Na,K)_{0-1}Ca_2(Mg,Fe,Al)_5[(Si,Al)_8O_{22}](OH,F)_2$

Biaxial - (+) Monoclinic

(IM 28, 34-37, 39, 40; MM 48)

Refractive Index Relief	Birefringence Interf. Color	2V	Extinction Elongation
x 1.61-1.70 y 1.62-1.73 z 1.63-1.73	0.02-0.03 low 2nd order	10-90	z^c 12-34 length slow

Color Pleochroism	Cleavage	Cleavage Angle	Hardness
green and brown x pale y dark z dark	{110} good	{110}^{1$\bar{1}$0} 56	6

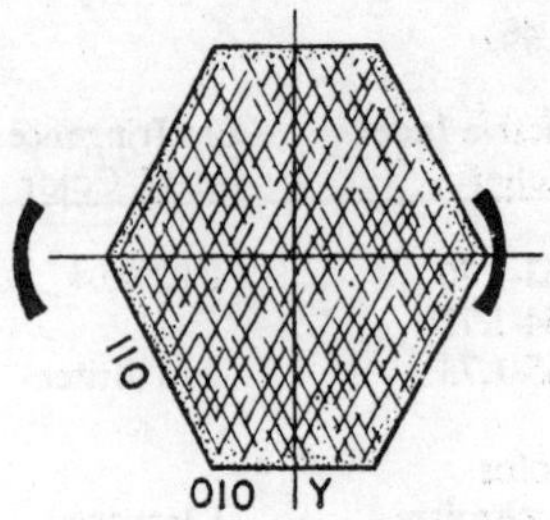

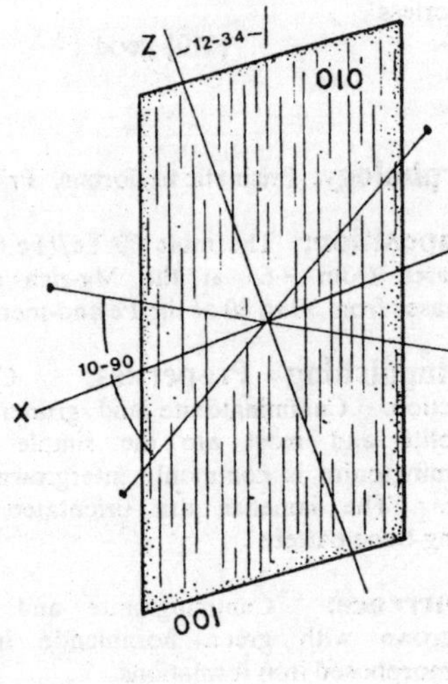

Morphology: Typically forms aggregates of prismatic grains in metamorphic rocks. Commonly forms phenocrysts which are 6-sided prisms in igneous rocks; these prisms tend to be elongated in felsic rocks and stubby in mafic rocks. Kaersutitic amphibole in intermediate to felsic rocks commonly grows as elongated prisms with an axial cavity filled with later-crystallizing minerals. In mafic and ultramafic alkaline rocks hornblende commonly crystallizes late and poikilitically encloses earlier-crystallizing minerals and thus has no crystal outline itself.

Composition: Common hornblende has compositions between tremolite-actinolite and pargasite-ferrohastingsite. This variation involves the substitution of $NaAl_3$ for $MgSi_2$ and $FeSi_2$, as seen by comparing the formulae of tremolite and pargasite, and ferroactinolite and ferrohastingsite:

Tremolite		Ca_2Mg_5	$[Si_8$		$O_{22}](OH,F)_2$
Subtract		Mg	Si_2		
Add	Na		Al	Al_2	
Pargasite					$NaCa_2Mg_4Al[Si_6Al_2O_{22}](OH,F)_2$

Ferroactinolite		Ca_2Fe_5	$[Si_8$		$O_{22}](OH,F)_2$
Subtract		Fe	Si_2		
Add	Na		Al	Al_2	
Ferrohastingsite					$NaCa_2Fe_4Al[Si_6Al_2O_{22}](OH,F)_2$

NaAl can replace Si to form an edenite-ferroedenite component: $NaCa_2(Mg,Fe)_5[AlSi_7O_{22}](OH,F)_2$,

and AlAl can replace MgSi to form a tschermakite-ferrotschermakite component: $Ca_2(Mg,Fe)_4Al[Al_2Si_6O_{22}](OH,F)_2$,

but the extent of these substitutions is usually not great. Refractive indices increase with increasing iron content but 2V decreases. Ferrohastingsite has a 2V of only 10, and because of the very strong absorption shown by this mineral the isogyres may not be clearly defined and the optic figure may appear almost uniaxial. The 2V increases to 90 and then becomes positive in pargasite.

Basaltic Hornblende resembles common hornblende, but much of the iron is Fe''', and O replaces some of the OH. It is formed in lava by oxidation following eruption. It has deep red and brown pleochroism.

Kaersutite is a yellow to deep reddish brown pleochroic amphibole that contains some ferric iron, but its chemical characteristic is a high Ti content. Distinguished from basaltic hornblende by dispersion--r>b in kaersutite, r<b in basaltic hornblende. Kaersutite is common in alkaline rocks. Its formula is $NaCa_2(Mg,Fe'',Fe''')_4Ti[Al_2Si_6O_{22}](OH,F)_2$.

Barkevikite is another deep reddish brown amphibole. It does not have as high an Fe''' content as basaltic hornblende, nor does it have as much Ti as kaersutite. It does, however, contain considerable Mn. Distinguished from other reddish brown amphiboles by 2V--barkevikite 40-50, kaersutite and basaltic hornblende 66-82. Like kaersutite, barkevikite is restricted to alkaline igneous rocks.

Distinguishing Properties: Common hornblende cannot be distinguished easily by optics from actinolite; hornblende does, however, have higher birefringence and larger 2V. Hornblende is distinguished from biotite in having two cleavage directions, inclined extinction, and lacking a birdseye texture.

Occurrence: Hornblende is one of the most common rock-forming minerals. One of the major minerals in metamorphosed mafic igneous rocks in the amphibolite facies. It is the main ferromagnesian mineral in many calcalkaline igneous rocks, and the kaersutitic and barkevikitic varieties are abundant in alkaline rocks.

GLAUCOPHANE $Na_2Mg_3Al_2[Si_8O_{22}](OH,F)_2$

Biaxial - Monoclinic

(MM 49-51)

Refractive Index Relief	Birefringence Interf. Color	2V	Extinction Elongation
x 1.61-1.66 y 1.62-1.67 z 1.63-1.67	0.01-0.02 1st order red	0-50	z^c 5 length slow

Color Pleochroism	Cleavage	Cleavage Angle	Hardness
blue-lavender x neutral y lavender z blue	{110} good	{110}^{1$\bar{1}$0} 58	6

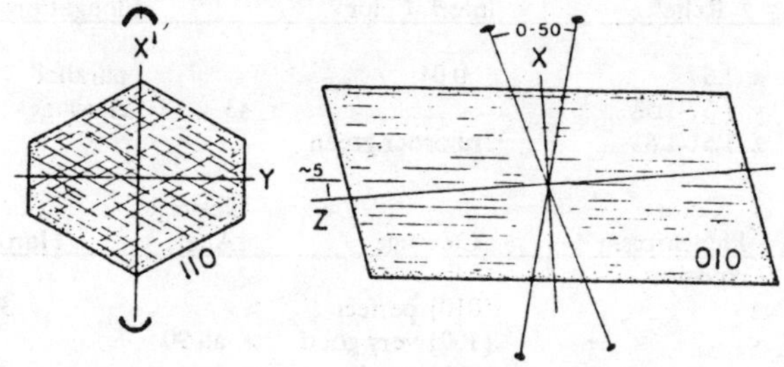

Morphology: Prismatic or columnar aggregates.

Composition: The formula of glaucophane can be derived from that of tremolite by substituting Na for Ca and maintaining charge balance by replacing Mg with Al. Typically some Fe substitutes for Mg.

Distinguishing Properties: Bluish pleochroism distinguishes glaucophane from most other minerals. Riebeckite is also blue, but it has a large 2V and is length fast.

Occurrence: Glaucophane is formed during regional metamorphism at low temperatures and high pressures. It is commonly associated with the mineral lawsonite; this assemblage characterizes the glaucophane schist facies. Because of the blue color of glaucophane, even in hand specimen, schists containing this mineral are commonly referred to as blueschists to contrast them with greenschists, which contain chlorite, epidote, or actinolite.

RIEBECKITE $Na_2Fe_3Fe_2[Si_8O_{22}](OH,F)_2$

Biaxial - (+) Monoclinic

(IM 38)

Refractive Index Relief	Birefringence Interf. Color	2V	Extinction Elongation
x 1.65-1.70 y 1.66-1.71 z 1.67-1.72	0.006-.016 gray but masked by blue color	40-90	x^c 5 length fast

Color Pleochroism	Cleavage	Cleavage Angle	Hardness
dark blue x deep blue y light blue z greenish	{110} good	{110}^{1$\bar{1}$0} 56	5

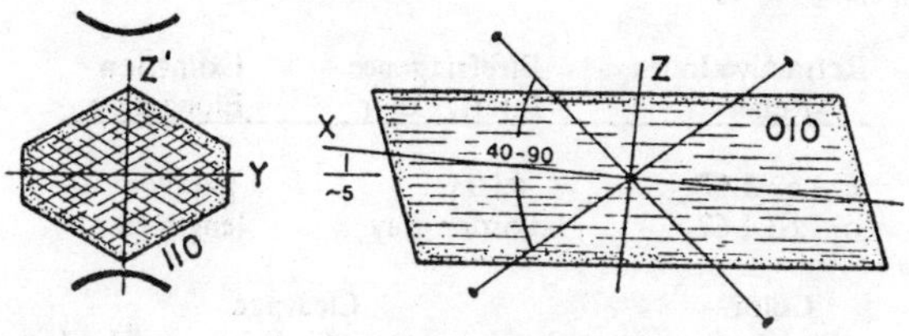

Morphology: Prismatic to fibrous.

Composition: Replacement of Ca by Na is coupled with substitution of Fe''' for Fe''.

Distinguishing Properties: The dark blue to green pleochroism of riebeckite distinguishes it from other minerals. Tourmaline is uniaxial and lacks cleavage. Glaucophane has paler colors, smaller 2V, and is length slow.

Occurrence: Riebeckite occurs in Na-rich felsic igneous rocks.

ANHYDRITE $CaSO_4$

Biaxial + | Orthorhombic

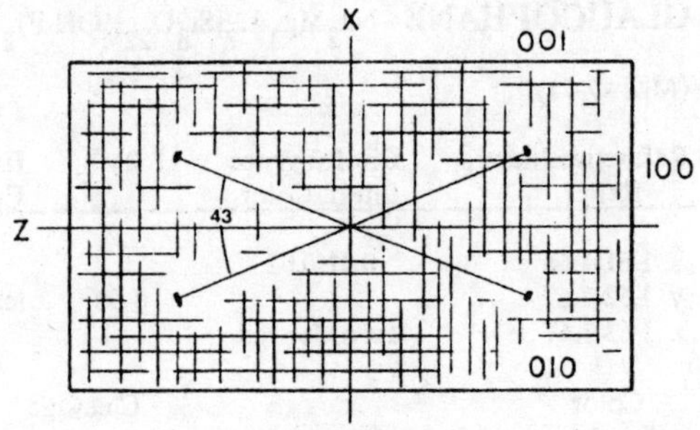

Refractive Index Relief	Birefringence Interf. Color	2V	Extinction Elongation
x 1.57	0.04		parallel
y 1.57-1.58		43	cleavages
z 1.61-1.62	3rd order green		

Color Pleochroism	Cleavage	Cleavage Angle	Hardness
colorless			
x	{010} perfect	>	3
y	{100} very good	> all 90	
z	{001} good	>	

Morphology: Generally forms anhedral aggregates, but tabular crystals do form in veins.

Composition: Small amounts of Sr and Ba may substitute for Ca.

Distinguishing Properties: The three prominent rectangular cleavages and high birefringence are characteristic.

Occurrence: The main mode of occurrence is in evaporite deposits. Near the Earth's surface it tends to alter to gypsum through the addition of water. It also forms in veins, but here too it tends to be altered or removed altogether, leaving only tabular molds.

APATITE $Ca_5(PO_4)_3(OH,F,Cl)$

(IM 73-76)

Uniaxial - | Hexagonal

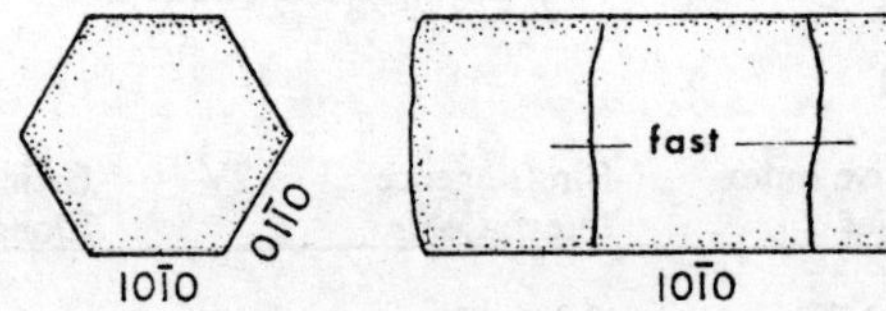

Refractive Index Relief	Birefringence Interf. Color	Extinction Elongation
e 1.62-1.67	0.003	parallel
o 1.63-1.67	1st order gray	length fast

Color Pleochroism	Cleavage	Cleavage Angle	Hardness
colorless			
e	none		5
o			

Morphology: Stubby hexagonal prisms in mafic igneous rocks, slender prisms in felsic igneous rocks, and anhedral grains in metamorphic rocks. Long slender prisms may have an axial cavity filled with later-crystallizing minerals.

Composition: The main variation is in the proportions of OH, F, and Cl; all possible extremes exist. The rare earth elements, in particular Ce, can substitute for Ca and commonly produce fluorescence when bombarded with electrons or ultraviolet radiation.

Distinguishing Properties: The low birefringent hexagonal prisms with high relief are characteristic of apatite. The mineral lacks cleavage, but prisms commonly have a widely spaced cross fracture. Sillimanite has similar cross fractures but has higher birefringence, is length slow and biaxial, and has square cross sections

Occurrence: The only common P-bearing mineral, and thus it is a ubiquitous accessory phase in igneous, metamorphic, and sedimentary rocks. Its abundance is determined simply by the abundance of P in the rock. Thus it is rare in most ultramafic rocks, relatively abundant in mafic and intermediate rocks, and less abundant in felsic ones.

BARITE $BaSO_4$

Biaxial + Orthorhombic

Refractive Index Relief	Birefringence Interf. Color	2V	Extinction Elongation
x 1.64 y 1.64 z 1.65	0.01 1st order yellow	 37	parallel {001} {001} slow

Color Pleochroism	Cleavage	Cleavage Angle	Hardness
colorless			
x	{001} perfect	{001}^{010} 90	3
y	{210} very good	{210}^{2$\bar{1}$0} 78	
z	{010} good		

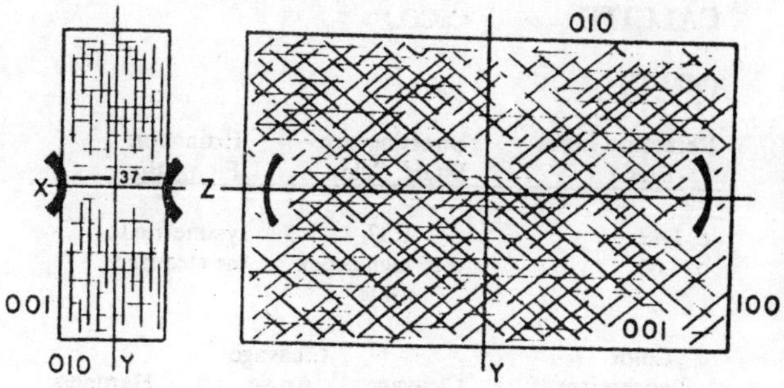

Morphology: Plate-like crystals flattened parallel to {001}.

Composition: Barite shows little variation in composition, but Sr can replace the Ba.

Distinguishing Properties: Cleavages distinguish barite from anhydrite and albite. Anhydrite also has much higher birefringence.

Occurrence: Commonly forms in veins with quartz and calcite. It occurs in the cement of some sandstones.

BERYL $Be_3Al_2[Si_6O_{18}]$

Uniaxial - Hexagonal

Refractive Index Relief	Birefringence Interf. Color	Extinction Elongation
e 1.56-1.60 o 1.56-1.60	0.004-0.009 1st order grey	parallel length fast

Color Pleochroism	Cleavage	Cleavage Angle	Hardness
colorless			
e -	{0001} poor		8
o -			

Morphology: Large hexagonal prisms with common termination by (0001).

Composition: Large holes in the ring structure of beryl permit the entry of large alkali ions; most beryl contains some alkalis. Small amounts of Cr account for the green color of the gem variety emerald.

Distinguishing Properties: The only mineral resembling beryl is apatite, but apatite has much higher refractive indices and lacks basal cleavage. Quartz is uniaxial positive.

Occurrence: Most beryl forms in pegmatites or late stage miarolitic cavities in granites and syenites. It also occurs in marbles and schists, the variety emerald being found typically in biotite schist.

CALCITE $CaCO_3$

Uniaxial - Trigonal

(MM 32)

Refractive Index Relief	Birefringence Interf. Color	Extinction Elongation
e 1.49 o 1.66	0.17 very high order white	symmetric to the cleavage

Color Pleochroism	Cleavage	Cleavage Angle	Hardness
colorless e but appears o cloudy	$\{10\bar{1}1\}$ perfect	75	3

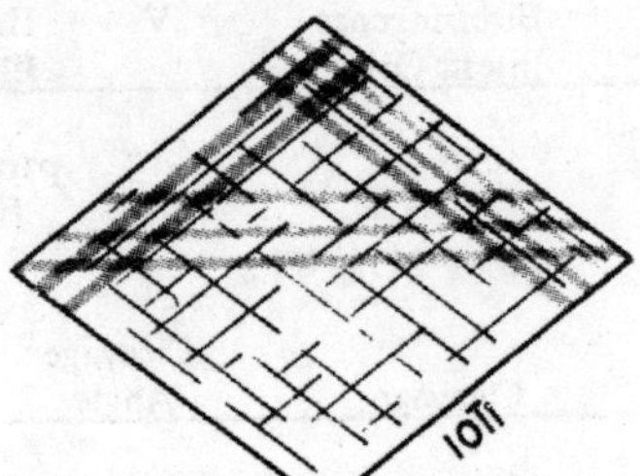

Morphology: Mostly anhedral except in veins and amygdules where it may form rhombohedrons or scalenohedrons. Lamellar twinning on $\{01\bar{1}2\}$ is particularly common.

Composition: Substitution of Mg, Fe, and Mn can occur. At low temperature, however, calcite is relatively pure.

Distinguishing Properties: Extreme birefringence distinguishes carbonates from most other minerals. Sphene has high birefringence but appears yellowish or brownish under plane light and is biaxial. Calcite and dolomite can be distinguished by the relation of the twin lamellae to the cleavage planes. In dolomite, twin lamellae parallel both the long and short diagonals of the cleavage rhombs; in calcite the twins parallel the long diagonal and the edges of the rhombs. The high-pressure polymorph of calcite, aragonite, also has extreme birefringence, but it is orthorhombic with a negative 2V of 18 and has only one poor cleavage, {010}, which has parallel extinction.

Occurrence: Calcite is a common mineral in sedimentary limestones, metamorphic marbles, hydrothermal veins, igneous carbonatite, and as a deuteric alteration product of mafic igneous rocks. The high pressure polymorph, aragonite, occurs in glaucophane schists, but it also occurs metastably at surface conditions when precipitated organically, as for example in shells. It also occurs metastably as fine needles filling vesicles and veins in basalt.

DOLOMITE $CaMg(CO_3)_2$

Uniaxial - Trigonal

Refractive Index Relief	Birefringence Interf. Color	Extinction Elongation
e 1.50 o 1.68	0.18 very high order white	symmetric to the cleavage

Color Pleochroism	Cleavage	Cleavage Angle	Hardness
colorless-gray e o	$\{10\bar{1}1\}$ perfect	75	4

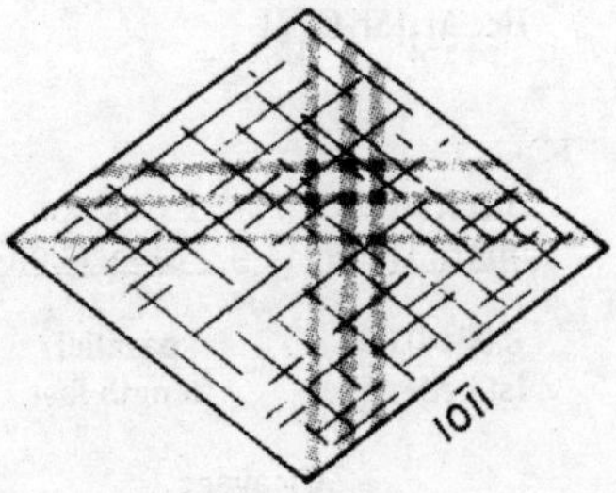

Morphology: Mostly anhedral but rhombs with curved crystal faces do occur.

Composition: Fe, Mn, and Ca can substitute for Mg. Varieties with more than 20% of the iron component are known as ankerite, and the iron end-member is siderite.

Distinguishing Properties: Dolomite has the extreme birefringence typical of carbonates. It is distinguished from calcite by having twinning parallel to both the short and long diagonals of the cleavage rhombs; calcite has twinning parallel to only the long diagonal and also edges of the cleavage rhombs.

Occurrence: Occurs in sedimentary dolostone, in metamorphic marbles at low to intermediate grades of metamorphism, in carbonatites, and as a vein mineral.

CHLORITE $(Mg,Al,Fe)_6[(Si,Al)_4O_{10}](OH)_8$ Biaxial + (-) Monoclinic

(MM 17-20)

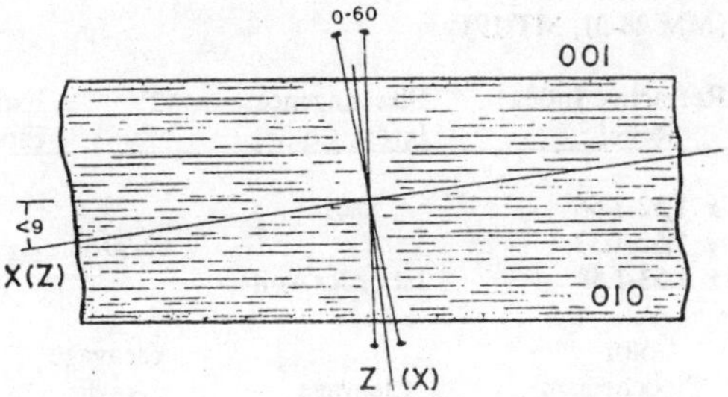

Refractive Index Relief	Birefringence Interf. Color	2V	Extinction Elongation
x 1.57-1.66 y 1.57-1.67 z 1.57-1.67	0-0.01 low 1st order anomalous	0-(60) most small	ext on {001} <9 most length fast

Color Pleochroism	Cleavage	Cleavage Angle	Hardness
green-colorless x pale green y pale green z colorless	{001} perfect		2

Morphology: Most forms micaceous plates, especially in metamorphic rocks, but when formed as an alteration of other minerals it may form a fine-grained aggregate..

Composition: Chlorite is a relatively magnesian mineral but it can contain considerable amounts of Fe, and its Al and Si contents also vary. In metamorphic rocks chlorite tends to be magnesian and aluminous. The chlorite formed as a deuteric alteration of ferromagnesian minerals in igneous rocks tends to be Fe-rich. Most chlorite is positive, but Fe- and Si-rich varieties are negative.

Distinguishing Properties: Resembles micas under plane light, but under crossed polars has very low birefringence. Most appears almost isotropic and many exhibit anomalous blue and brown interference colors. Dispersion is strong (r<b). Serpentine has lower refractive indices, lacks pleochroism, and has weak (r>b) to no dispersion.

Occurrence: An important constituent of many low grade metamorphic rocks, and the name greensc hist facies is derived from the color of this mineral in hand specimen. It occurs in both meta-pelites and meta-mafic igneous rocks. Chlorite is also a common alteration product of ferromagnesian minerals formed during the cooling of igneous rocks. It forms hydrothermally in mafic igneous rocks and is commonly associated with hydrothermal ore deposits.

CHLORITOID $(Fe^{2+},Mg,Mn)_2(Al,Fe^{3+})Al_3O_2[SiO_4]_2(OH)_4$ Biaxial + Monoclinic

(MM 26,27)

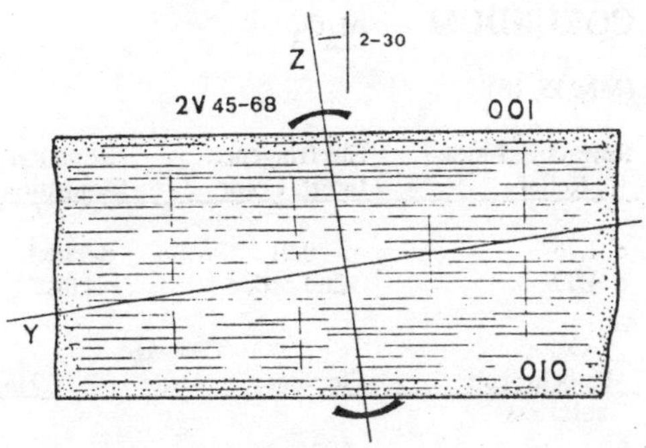

Refractive Index Relief	Birefringence Interf. Color	2V	Extinction Elongation
x 1.71-1.73 y 1.72-1.73 z 1.72-1.74	0.01-0.02 low 1st order anomalous	45-68	ext on {001} <30 {001} fast

Color Pleochroism	Cleavage	Cleavage Angle	Hardness
green and gray x gray-green y gray-blue z colorless	{001} good		6

Morphology: Forms tabular crystals that may exhibit hourglass zoning. It typically grows as porphyroblasts in meta-pelites. Lamellar twinninig on {001} is very common.

Composition: Chloritoid is restricted to Fe-rich compositions.

Distinguishing Properties: Pleochroism in shades of green, gray, and blue are distinctive, especially when emphasised by the twinning. Resembles chlorite slightly, but chloritoid's refractive indices are much higher, it is distinctly biaxial whereas most chlorite appears uniaxial, and it exhibits cross fractures. Chloritoid is also much harder than chlorite.

Occurrence: Forms in Fe-rich pelitic rocks at low to medium grades of regional metamorphism.

CORDIERITE $(Mg,Fe)_2Al_3[Si_5AlO_{18}]$

Biaxial - Orthorhombic

(MM 28-31; MT 19)

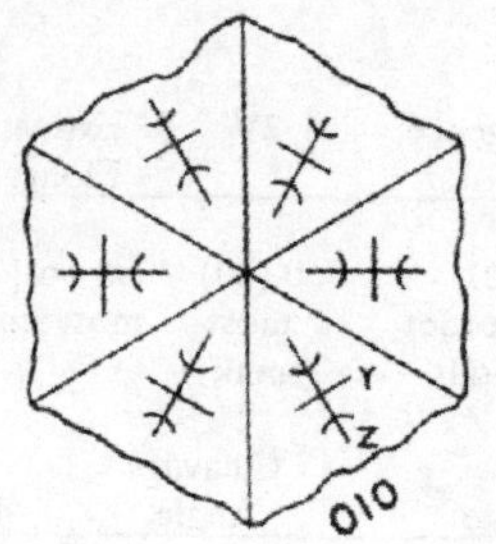

Refractive Index Relief	Birefringence Interf. Color	2V	Extinction Elongation
x 1.52-1.56	0.01		
y 1.52-1.57		60-90	
z 1.53-1.58	1st order white		

Color Pleochroism	Cleavage	Cleavage Angle	Hardness
colorless			
x	{010} poor		7
y			
z			

Morphology: Occurs as anhedral grains, ovoid-shaped porphyroblasts, and pseudohexagonal crystals. The hexagonal outline results from twinning, with each member of the twin constituting a triangular sector of the hexagon. Lamellar twinning similar to that in plagioclase is also common.

Composition: Although some substitution of Fe can take place, cordierites are restricted to Mg-rich compositions. Water can be present in large channels in the structure at low temperatures but is absent from cordierite formed at high temperature.

Distinguishing Properties: Only common ferromagnesian mineral that can be confused with feldspar. Its refractive indices and lamellar twinning appear identical to those in plagioclase. The triangular sector twins are characteristic of cordierite, and if it includes zircon grains, pleochroic haloes may be present (they would not be in plagioclase). Cordierite typically contains many small opaque inclusions. It alters readily along irregular fractures and grain boundaries to broad feathery zones of micaceous material (pinite). Quartz is distinguished by its uniaxial figure and lack of twinning.

Occurrence: A common metamorphic mineral in pelitic rocks. Its occurrence is favored by low pressures or high temperatures. It is therefore more common in rocks affected by contact metamorphism or regional metamorphism at low pressures. It does, however, remain stable to high pressures and occurs in granulite facies rocks. Rare occurrences of cordierite in igneous rocks may result from the assimilation of aluminous sediments.

CORUNDUM Al_2O_3

Uniaxial - Trigonal

(MM 15, 16)

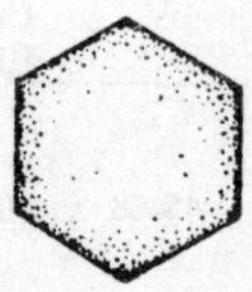

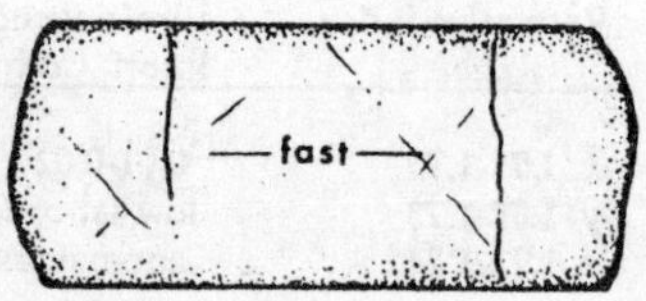

Refractive Index Relief	Birefringence Interf. Color	Extinction Elongation
e 1.76	0.01	parallel
o 1.77	2nd order	c is fast

Color Pleochroism	Cleavage	Cleavage Angle	Hardness
colorless			
e	none		9
o			

Morphology: Normally forms euhedral crystals that are either six sided prisms or plates. Corundum is so hard that its thickness may be greater than that of the rest of the thin section; thus its interference colors may be higher than would be expected from the mineral's low birefringence.

Composition: Solution of small amounts of Fe and Ti produces the variety known as sapphire, whereas small amounts of Cr produce ruby.

Distinguishing Properties: High relief, moderately low interference colors, uniaxial figure, and lamellar twinning on $\{10\bar{1}1\}$ characterize corundum.

Occurrence: Produced by extremely high grade contact metamorphism of pelitic rocks where it may be associated with spinel to form emery. Found in some syenites and nepheline syenites, and in mantle and lower crustal nodules in kimberlite.

EPIDOTE $Ca_2Fe^{3+}Al_2O[Si_2O_7][SiO_4](OH)$ Biaxial - Monoclinic

(MM 41-43)

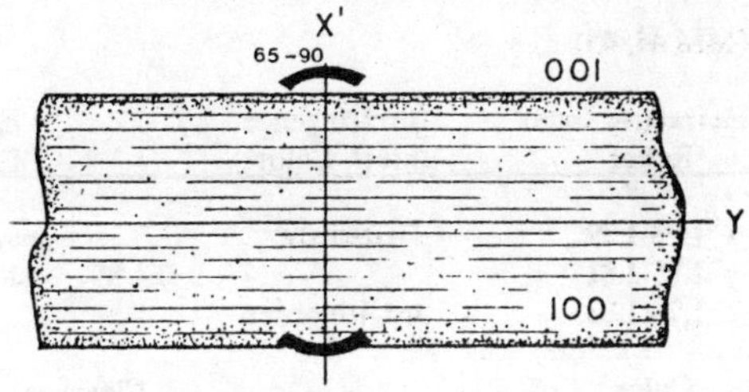

Refractive Index Relief	Birefringence Interf. Color	2V	Extinction Elongation
x 1.72-1.75	0.02-0.05		parallel length
y 1.72-1.78	bright	65-90	slow and fast
z 1.73-1.80	2nd-3rd order		

Color Pleochroism	Cleavage	Cleavage Angle	Hardness
yellowish green			
x pale yel. green	{001} perfect		6
y greenish yellow			
z yellowish green			

Morphology: Most forms granular aggregates, but some forms crystals elongated along their b axis.

Composition: Substitution of Al for Fe''' causes epidote to grade into clinozoisite (see below); this decreases its birefringence and refractive indices.

Distinguishing Properties: Very bright and variable interference colors are characteristic. Because epidote crystals are elongated parallel to crystallographic b, crystals are both length slow and length fast. Distinguished from augite in sections showing a single cleavage parallel to the length of the grain by having the optic plane perpendicular to the cleavage.

Occurrence: An extremely common mineral in medium grades of regional metamorphism, especially in meta-mafic igneous rocks. It also occurs as a deuteric mineral and fills vesicles and fractures in basalts.

CLINOZOISITE $Ca_2AlAl_2O[Si_2O_7][SiO_4](OH)$ Biaxial + Monoclinic

(MM 43)

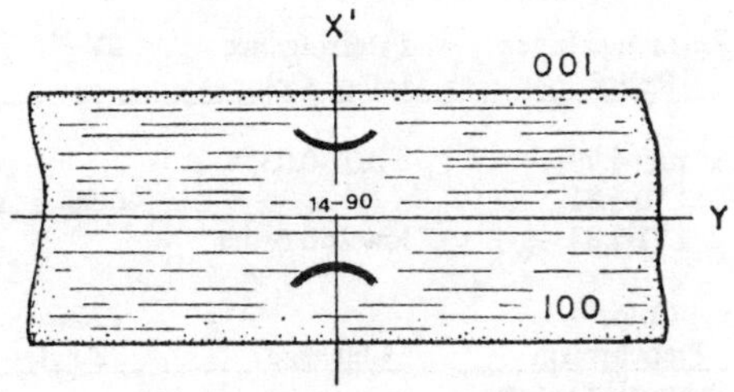

Refractive Index Relief	Birefringence Interf. Color	2V	Extinction Elongation
x 1.67-1.71	0.005-0.015		parallel length
y 1.67-1.72	anomalous	14-90	slow and fast
z 1.69-1.73	low 1st order		

Color Pleochroism	Cleavage	Cleavage Angle	Hardness
colorless			
x -	{001} perfect		6.5
y -			
z -			

Morphology: Commonly forms crystals elongated parallel to crystallographic b, but also forms granular aggregates.

Composition: An aluminum-rich, iron-poor end member of the epidote series. At the extreme iron-poor end of the series orthohombic zoisite forms. Refractive indices, birefringence, and 2V of clinozoisites increase with increasing iron content.

Distinguishing Properties: One of a small group of minerals (+ melilite, vesuvianite, chlorite) that exhibits anomalous blue interference colors. It differs from melilite and vesuvianite in being biaxial, and from chlorite by its much higher refractive indices. Clinozoisite commonly occurs in grains that are zoned to more iron-rich compositions and may therefore show considerable variation in birefringence.

Occurrence: The occurrence of clinozoisite is essentially the same as that of epidote, occurring in amphibolites. It is particularly associated with the alteration of calcic plagioclase.

PIEMONTITE $Ca_2(Mn^{+3},Fe^{3+},Al)_3O[Si_2O_7][SiO_4](OH)$

Biaxial + Monoclinic

(MM 44, 45)

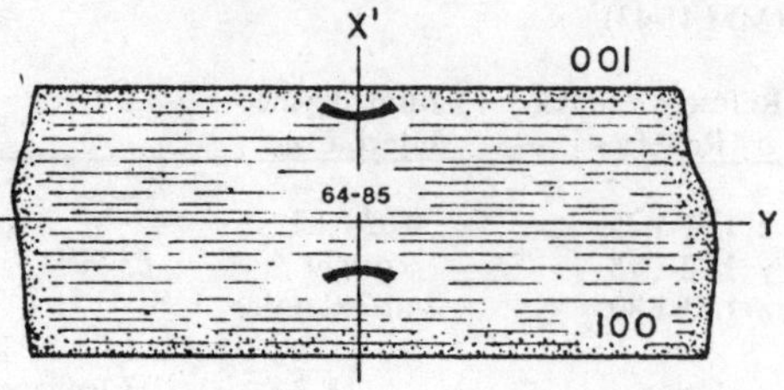

Refractive Index Relief	Birefringence Interf. Color	2V	Extinction Elongation
x 1.73-1.79 y 1.75-1.81 z 1.76-1.83	0.03-0.09 3rd-4th order	 64-85	parallel length slow and fast

Color Pleochroism	Cleavage	Cleavage Angle	Hardness
x yellow y amethyst z red	{001} perfect		6

Morphology: Commonly forms prismatic crystals elongated parallel to crystallographic b.

Composition: Forms a solid solution series with normal epidote.

Distinguishing Properties: The striking pleochroism distinguishes it from other minerals.

Occurrence: Forms over the same range of conditions as other epidotes, but is restricted to metamorphosed manganiferous rocks, such as cherts, which form piemontite- spessartine-bearing quartzite.

ALLANITE $(Ca,Ce)_2(Fe^{2+},Fe^{3+})Al_2O[Si_2O_7][SiO_4](OH)$

Biaxial -+ Monoclinic

Refractive Index Relief	Birefringence Interf. Color	2V	Extinction Elongation
x 1.69-1.79 y 1.70-1.81 z 1.71-1.83	0.01-0.03 low 2nd order	 40-90	parallel length slow and fast

Color Pleochroism	Cleavage	Cleavage Angle	Hardness
brownish yellow x light brown y yellowish brown z dark brown	 {001} poor		 6

Morphology: Forms stubby prisms which are commonly very cracked due to the metamict nature of the mineral.

Composition: This Ce-bearing epidote commonly contains other rare earth elements and Th and U which, on undergoing radioactive decay, destroy the structure of the mineral, making it metamict.

Distinguishing Properties: Optical properties may be difficult to obtain if the mineral is strongly metamict. Metamict varieties may appear isotropic. Pleochroic haloes may be produced in juxtaposed biotite. Distinguished from brown amphibole by a single cleavage, against which the extinction is parallel in elongated grains.

Occurrence: Common accessory in granites, syenites,and pegmatites; it is found alro in some gneisses and contact skarn deposits.

Feldspars are the most common rock-forming minerals in crustal rocks. They consist of three essential components, a Ca end-member, Anorthite (An), a Na end-member, Albite (Ab), and a K end-member, Orthoclase (Or). At high temperatures, such as those at which most volcanic rocks form, complete solid solution exists between anorthite and albite to form the plagioclase feldspars and between albite and orthoclase to form the alkali feldspars; only very limited solid solution, however, occurs between anorthite and orthoclase. At lower temperatures a wide miscibility gap separates Na- and K-rich alkali feldspars, and smaller miscibility gaps are present in the plagioclase series.

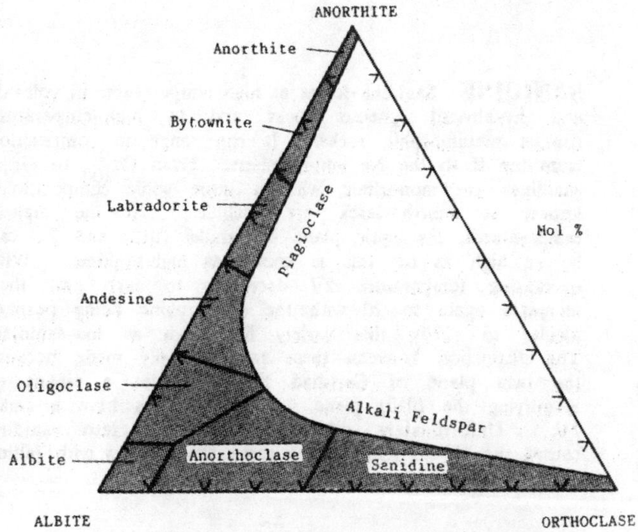

Compositions of feldspars are expressed in terms of mole percentages of the three essential components, and are written as $An_xAb_yOr_z$, where x, y, and z indicate the mole percentages of anorthite, albite, and orthoclase respectively. Because the amount of orthoclase component in plagioclases, especially calcic ones, is so small, plagioclase is commonly expressed in terms only of its anorthite-albite content (recalculated to 100%), which is written as $An_{\%}$.

The structure of feldspars consists of $(Si,Al)O_4$ tetrahedra linked together by sharing of all oxygens into a three-dimensional network in which large holes are filled with Ca, Na, or K ions. At high temperatures Al and Si fill the tetrahedral sites randomly. These sites are not all identical, and at lower temperatures site preferences develop which cause the structures to change slightly. The result is that the optical properties of feldspars are not only a function of composition but also of what is referred to as their structural state. Although this complicates the identification of feldspars, it can provide information on their temperatures of formation. This is particularly true in the alkali feldspars, where the degree of ordering of Al and Si on tetrahedral sites causes readily observable optical differences.

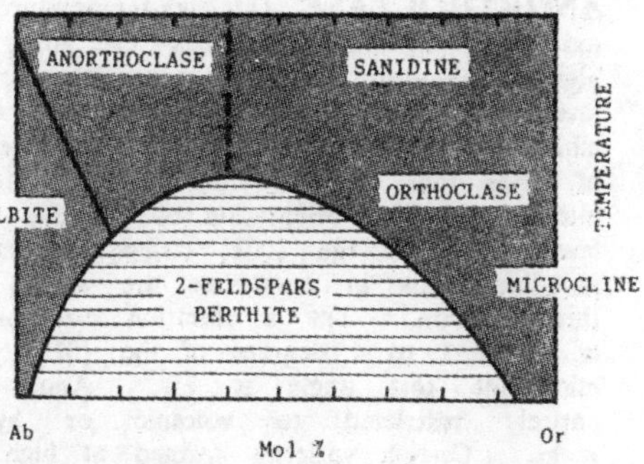

Schematic representation of relative temperatures at which various alkali feldspars form.

ALKALI FELDSPAR

$(K,Na)[AlSi_3O_8]$ Biaxial - Monoclinic, Triclinic

Refractive Index Relief	Birefringence Interf. Color	2V	Cleavage Plane	Angle	Hardness
K - Na					
x 1.518-1.529	0.007		perfect {010}	~90	6
y 1.522-1.533		0-80	perfect {001}		
z 1.522-1.539	1st order gray		poor {100}	~116	

General Optical Properties Alkali feldspars are colorless in thin section, but they may appear a cloudy gray or brown due to alteration to clay minerals. They may form euhedral rectangular crystals especially as phenocrysts in igneous rocks. They also occur as anhedral grains in many granites and high-grade metamorphic rocks. Two prominent cleavages intersect at about 90°. Twinning is common. In most igneous alkali feldspars a Carlsbad twin is present. In the triclinic feldspars albite and pericline twins can also be present. In general, the 2V of the alkali feldspars increases with increasing ordering of the Al and Si in the tetrahedral sites. Thus, most high-temperature alkali feldspar formed in volcanic rocks and cooled rapidly preserve the disordered Al and Si, and the 2V is small. In plutonic igneous rocks and metamorphic rocks where cooling rates are slow, ordering has more chance to occur, and the 2V becomes larger. Authigenically-formed alkali feldspar in sedimentary rocks has a large 2V. With falling temperature homogeneous alkali feldspars may exsolve into K- and Na-rich phases. This exsolution produces lenses of one feldspar in the other and is referred to as perthite or microperthite depending on whether a microscope is necessary to observe the phenomenon. Exsolution may develop in rapidly-cooled feldspars on such a fine scale that the two phases can be detected only by x-rays; these apparently optically homogeneous feldspars are known as cryptoperthites.

SANIDINE Sanidine forms at high temperatures in volcanic and hypabyssal igneous rocks and in high-temperature contact metamorphic rocks. It can range in composition from the K to the Na end-members. From Or_{100} to Or_{37}, sanidines are monoclinic, whereas more sodic compositions, known as anorthoclases, are triclinic. At the highest temperatures, the optic plane is parallel (010) and 2V can be as high as 62; this is known as high-sanidine. With decreasing temperature 2V decreases to zero and then increases again to 20 with the optic plane being perpendicular to (010); this variety is known as low-sanidine. The distinction between these two is easily made because the twin plane of Carlsbad twins provides a means of identifying the (010) plane. Most sanidines have a small 2V. Unfortunately, substitution of albite into sanidine causes the 2V to increase the same way it does with falling temperature.

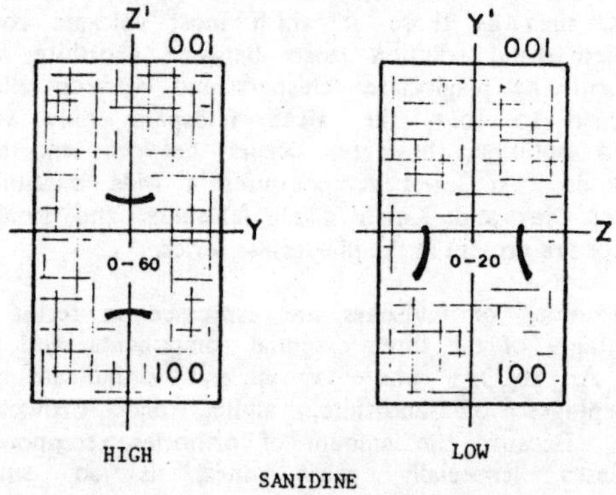

ANORTHOCLASE This high-temperature feldspar is restricted to compositions between $Or_{37}Ab_{63}$ and albite. At high temperature it is monoclinic, but it invariably inverts to triclinic on cooling and as a result develops albite and pericline twins. This produces a grid pattern of twins that resembles but is much finer than that in microcline. The orientation of pericline twins is within a few degrees of the {001} cleavage direction, whereas the pericline twins in microcline are almost perpendicular to this direction. The x vibration direction in anorthoclase is within 1 to 4 degrees of the {010} plane, whereas in microcline this angle is 15. Anorthoclase is almost entirely restricted to volcanic or hypabyssal igneous rocks. Ca-rich varieties formed at high temperature and cooled slowly unmix to form mesoperthite, which has roughly equal proportions of oligoclase and orthoclase.

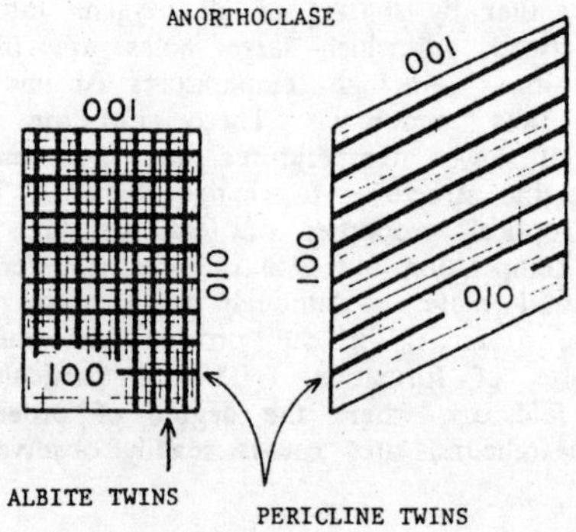

ORTHOCLASE (IM 56, 57) Orthoclase is monoclinic but has a larger 2V than sanidine (55-80) as a result of partial ordering in the tetrahedral sites. In igneous rocks, crystals commonly have a Carlsbad twin and contain exsolution lamellae of albite. Orthoclase is a common constituent of felsic plutonic igneous rocks and granulite facies metamorphic rocks.

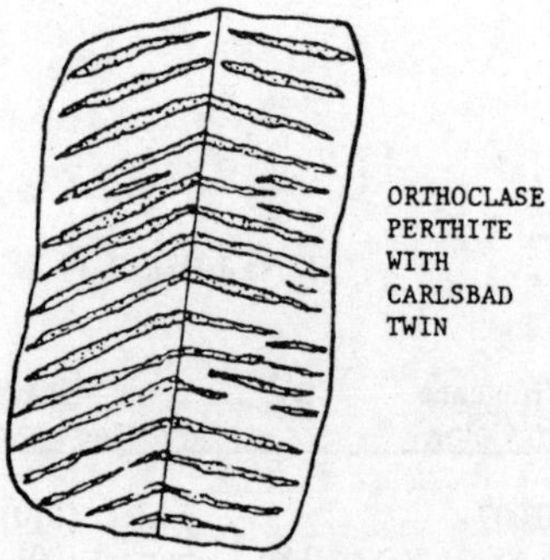

MICROCLINE (IM 58) With sufficient ordering of the Al and Si in tetrahedral sites, alkali feldspar becomes triclinic and is known as microcline. Almost all microcline exhibits a fine grid twinning produced by albite and pericline twins. This twinning is developed on inversion from the higher temperature monoclinic form. Some rare microcline found in sedimentary rocks grows initially in the triclinic form and lacks grid twinning. The 2V of microcline is near 80, but twinning makes its determination difficult.

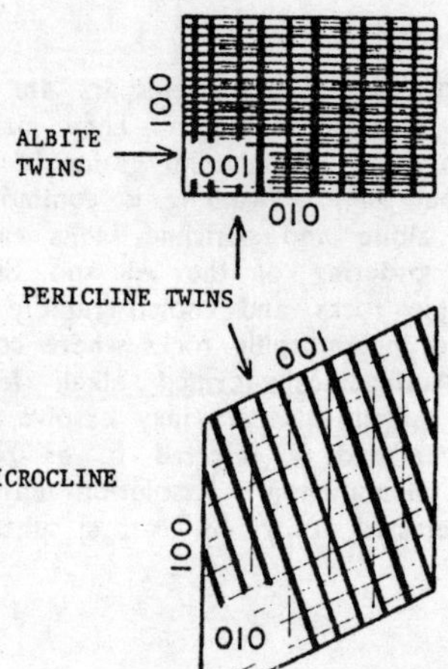

PLAGIOCLASE $Na[AlSi_3O_8]$ - $Ca[Al_2Si_2O_8]$

Biaxial + (-)

Triclinic

(IM 44-55)

Refractive Index Relief	Birefringence Interf. Color	2V	Cleavage Plane	Cleavage Angle	Hardness
Na - Ca	Na - Ca				
x 1.527-1.577	0.010-0.013		perfect {001}	~90	6
y 1.531-1.585	1st order	76-90	good {010}		
z 1.538-1.590	white-yellow				

General Optical Properties

Plagioclase is the most common rock-forming mineral in the Earth's crust. Composition varies widely but can be determined easily. The determination is one of the important steps in identifying and classifying rocks.

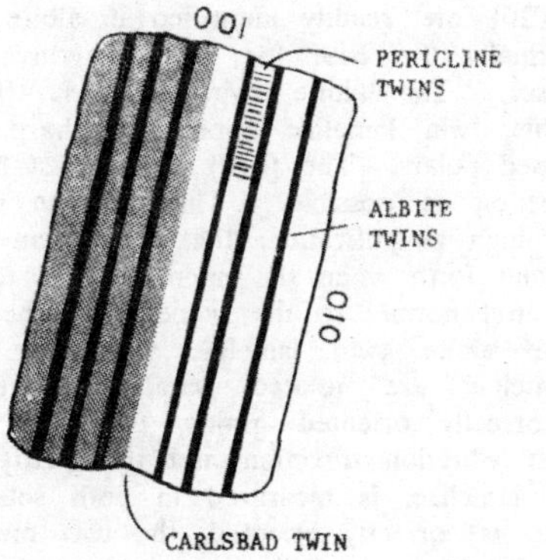

Despite formation over a wide range of conditions and in rocks of very different compositions, plagioclase has a distinct and easily identified appearance. In igneous rocks, crystals tend to form laths flattened on (010), but anhedral forms also occur. Twinning is common, with crystals typically having a single Carlsbad twin and multiple albite twins. In metamorphic rocks, most plagioclase is anhedral, and twinning is much less common; care must therefore be taken not to confuse plagioclase with quartz. Interference colors range up to first order white for albite and up to a pale first order yellow for anorthite. The relief is almost nil, with Ab-rich varieties being slightly negative and An-rich ones positive.

Because plagioclase belongs to a continuous solid solution series, compositional variations within crystals are common. Indeed, most igneous plagioclase exhibits some compositional zoning, and in rapidly-cooled rocks and ones of dioritic and granodioritic compositions zoning may be prominent. Because plagioclase crystallizing from melts decreases in Ca content with falling temperature, crystals that range in composition from Ca-rich cores to Na-rich rims are said to be normally zoned, whereas those that are more calcic on the rim are reversely zoned. Some crystals exhibit zoning which oscillates between slightly different compositions. This type, which resembles the growth bands in a tree, is known as oscillatory zoning. Although zoning is less common in metamorphic plagioclase, normal and reverse zoning is found, but not oscillatory zoning.

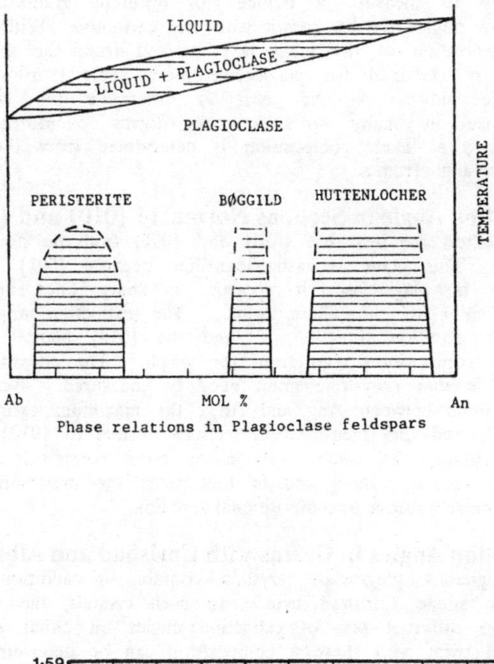

Phase relations in Plagioclase feldspars

The temperatures at which low-grade metamorphic rocks form are below the miscibility gap at the Na-rich end of the plagioclase series. Consequently, plagioclase within this compositional range cannot form in metamorphic rocks. Moreover, plagioclase on the Ca side of the miscibility gap reacts to form epidote at low metamorphic grades. The composition of plagioclase in low-grade metamorphic rocks is thus restricted to the Ab-rich end of the series. More calcic plagioclase does not form until the amphibolite facies is reached. Igneous plagioclase form at temperatures above the peristerite field and on slow cooling can develop fine exsolution lamellae, which may give the crystals an iridescent color. Fine exsolution lamellae can form in two other ranges of plagioclase composition, with that formed near An_{50} producing the striking blue and green iridescence known as labradorescence.

Refractive indices of plagioclase increase linearly with increasing An content. Albitic plagioclase has negative relief with respect to Canada balsam (R.I. = 1.537); be wary of epoxies with higher refractive indices. Plagioclase more calcic than about An_{10} has positive relief. Plagioclase more calcic than about An_{30} has greater refractive indices than those of quartz. This is a useful comparison because several of the determinative methods described below may give two possible compositions, one greater than and the other less than about An_{30}.

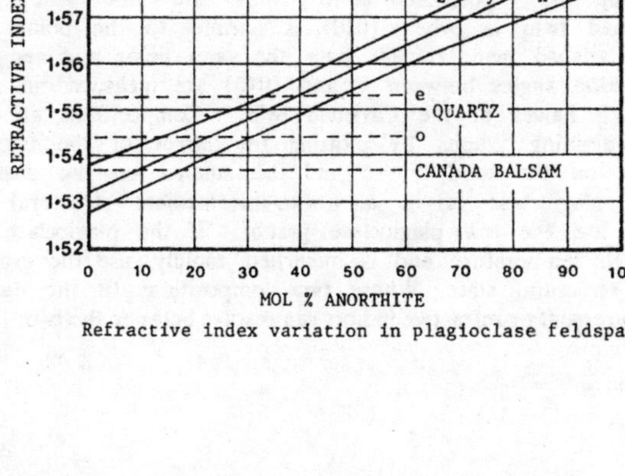

Refractive index variation in plagioclase feldspars

Because plagioclase is triclinic the orientation of the indicatrix has no crystallographic restrictions; the indicatrix rotates approximately one degree for each percent change in the An content, thus providing a simple means of determining compositions. The determination involves measuring the orientation of the indicatrix with respect to crystallographic axes. With a universal stage, where the precise orientation of the indicatrix can be measured, this is a simple task. On the flat stage of the microscope, however, we are limited to measuring random sections through the indicatrix and having to relate these to crystallographic directions that are indicated by known twin orientations or prominent cleavages. Three of the most commonly used methods are described below.

Maximum Extinction Angle in Sections Normal to {010}

Sections normal to {010} are readily identified if albite twins are present, which is normally the case for most igneous and some metamorphic plagioclase. The albite twin plane is {010}. In sections normal to this, twin lamellae appear as sharp dark and light bands under crossed polars. The {010} plane must be as close to normal to the section as possible. This can be checked in several ways. Under high magnification, the albite twin plane will appear to shift back and forth when the microscope is focussed up and down if {010} is not normal to the section. If the section is correctly oriented, the albite twin lamellae will have a similar color when the lamellae are rotated parallel to the polars (//crosshairs). In correctly oriented grains, then, the extinction angle between the fast vibration direction and the {010} plane, as indicated by the twin lamellae, is measured in both sets of albite lamellae. If the grain is correctly oriented, the two measurements should be identical. If they are close, an average can be taken. If they differ by more than 5° the grain is too misoriented to be used. The fast vibration direction obtained in this way is only by chance the x direction in the indicatrix. But only sections that have the x vibration direction parallel to the stage of the microscope will give the maximum extinction angle; any other section will give a component of this angle. It is therefore necessary to measure a number of different grains before the maximum angle can be known with any certainty. With this angle, the composition of the plagioclase is read from the graph. Two values are obtained for plagioclase more sodic than An_{40}, but refractive indices can be used to determine the correct one. Plagioclase in many volcanic rocks forms microlites elongated parallel to a. Their composition is determined from the maximum extinction angle from a.

Extinction angles to x' for various sections through plagioclase crystals.

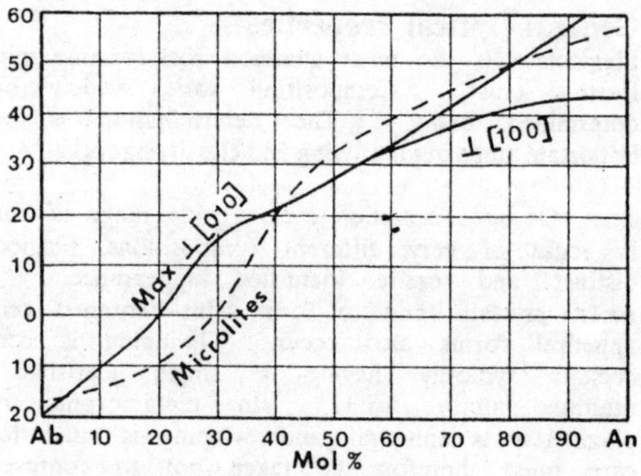

Extinction Angle in Sections Normal to {010} and {001}

This method requires that {010} and {001} both be normal to the section. This plane is easily identified because {001} is the best cleavage in plagioclase; it appears as sharp lines approximately perpendicular to the albite twins. The extinction angle between the fast vibration direction, x', and the {010} plane is measured and the composition read from the graph. The advantage of this method is that only one grain need be measured. Note that for compositions between An_0 and An_{60} the maximum extinction angle on {010} and that measured in sections normal to {010} and {001} are essentially the same. At more calcic compositions, however, the two curves deviate, and in this range the maximum extinction angle is more sensitive to compositional variation.

Extinction Angles in Grains with Carlsbad and Albite Twins

Many igneous plagioclase crystals contain, in addition to albite twins, a single Carlsbad twin. In such crystals, the albite twins give two different sets of extinction angles on either side of the Carlsbad twin; with these a composition can be determined directly. As in the other methods, the {010} plane must be vertical. Crystals with Carlsbad and albite twins are illustrated in Figs. IM 48-53. In the 45° position albite twins are not visible, but the Carlsbad twin is; when {010} is parallel to the polars the albite and Carlsbad twin lamellae have the same color and are not visible. Extinction angles between x' and {010} are measured in albite twins in both halves of the Carlsbad twin. Compositions are read from the adjoining graphs by plotting the larger of the two extinction angles on the solid curves and the smaller on the dashed curves. Most plagioclase is in a low-to-intermediate structural state; for these use the low plagioclase graph. If the plagioclase is formed at high temperature and is quenched rapidly, use the graph for the high structural state. Where two compositions fit the data, another measurement or refractive indices can resolve between the two.

Extinction angles between x' and (010) in zone ⊥(010) in plagioclase crystals exhibiting a Carlsbad twin. The larger extinction angle of the Carlsbad twin is given by the solid lines, and the smaller by the dashed lines. Extinction in sections ⊥[100] and //[100] are also given. The two graphs are for plagioclase in the low and high structural states. (After Tobi and Kroll, 1975; published with permission of the American Journal of Science)

LOW (PLUTONIC)

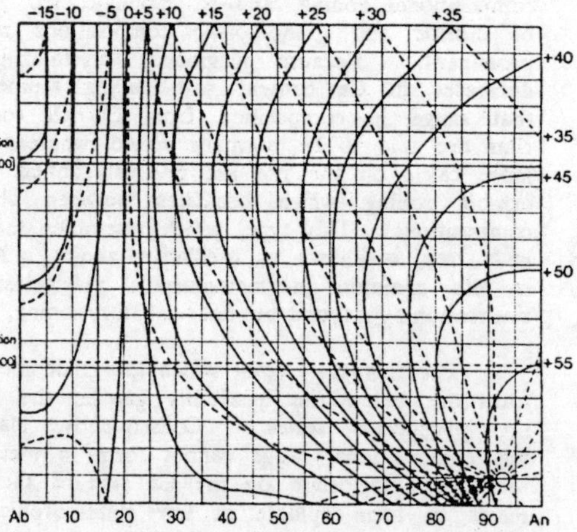

HIGH (VOLCANIC)

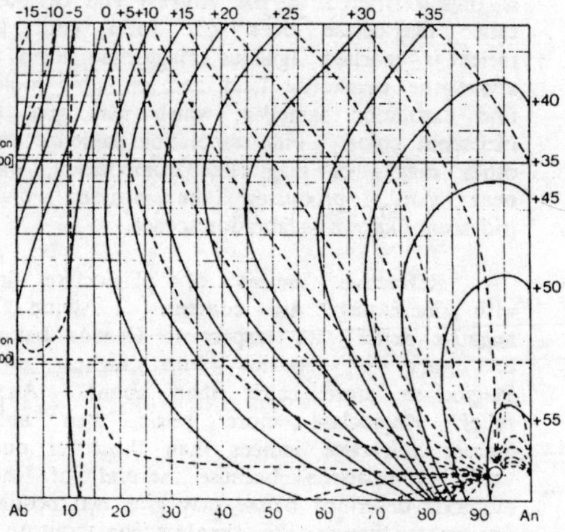

FLUORITE CaF_2 — Isotropic — Cubic

Refractive Index	Color	Birefringence	Cleavage	Hardness
1.434	Colorless Purple	Isotropic	{111} perfect	4

Morphology: In veins, fluorite may form cubes and octahedra, but in igneous rocks it crystallizes late and its shape is determined by the morphology of surrounding earlier-formed minerals.

Composition: Essentially pure calcium fluoride.

Distinguishing Properties: Has the lowest refractive index of any of the common minerals. It has strong negative relief. It is isotropic and commonly has purple spots and bands, especially in contact with radioactive minerals.

Occurrence: Common in hydrothermal veins and as a late-crystallizing accessory in some granites, syenites, and nepheline syenites.

GARNET $(Mg,Fe,Ca)_3Al_2[SiO_4]_3$ — Isotropic — Cubic

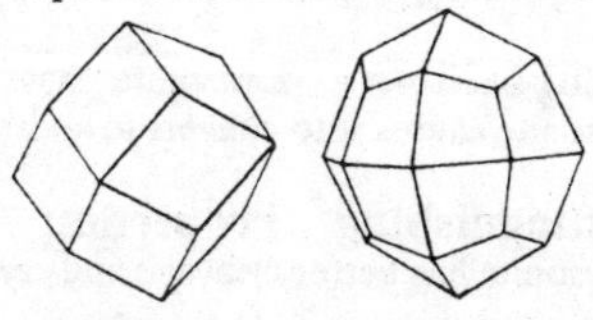

(MM 21-23)

Refractive Index	Color	Birefringence	Cleavage	Hardness
1.72-1.89	red, pink brown	Ca varieties anomalous	none	7

Morphology: Six- and eight-sided sections through euhedral crystals are common especially in schists. Anhedral grains are also common, and they may enclose many small grains of quartz.

Composition: Shows extensive solid solution between six common end-members. In general they fall into two groups depending on the presence of Ca, which in turn depends on the composition of the rock.

		Color in hand specimen	
Pyrope	$Mg_3Al_2[SiO_4]_3$	Pale pink	
Almandine	$Fe''_3Al_2[SiO_4]_3$	Deep red	
Spessartine	$Mn_3Al_2[SiO_4]_3$	Red	weak birefringence
Grossularite	$Ca_3Al_2[SiO_4]_3$	Honey-brown	anomalous birefringence
Andradite	$Ca_3(Fe'''Ti)_2[SiO_4]_3$	Yellow-brown	anomalous birefringence
Uvarovite	$Ca_3Cr_2[SiO_4]_3$	Green	anomalous birefringence

Distinguishing Properties: The only other common high refractive index isotropic mineral is spinel, which occurs as octahedra or anhedral grains. Common spinels are green, except for those containing Cr which are dark brown to opaque. The Ca-rich garnets commonly have anomalous low interference colors which may allow faint sector twins to be visible. Spessartine may also exhibit weak birefringence.

Occurrence: One of the most common metamorphic minerals and is found in a wide range of contact and regional metamorphic rocks. Garnets rich in the almandine component (with minor spessartine and pyrope) are common in intermediate- to high-grade regionally metamorphosed pelitic and mafic igneous rocks. The cores of these crystals tend to be Mn-rich. With increasing metamorphic grade and particularly with increasing pressure, the pyrope content increases. Ca-rich garnets form in calcareous rocks both of regional and contact metamorphic origin. Granites contaminated with aluminous rocks may crystallize almandine, and some nepheline syenites contain a Ti-rich andradite known as melanite. Mantle-derived nodules in kimberlites commonly contain pyrope-rich garnet.

GYPSUM $CaSO_4.2H_2O$ — Biaxial + — Monoclinic

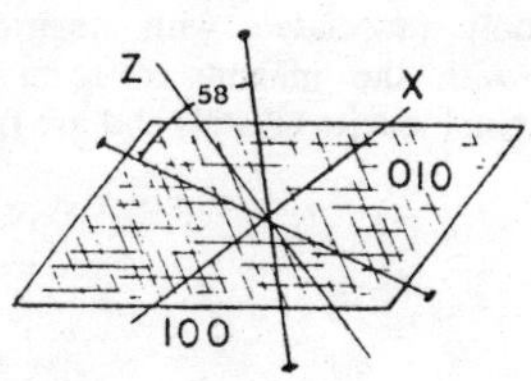

Refractive Index	Birefringence Interf. Color	2V	Cleavage Plane	Hardness
x 1.52	0.01	58	{010} perfect	2
y 1.53			{100} fair	
z 1.53	1st order white		{011} fair	

Morphology: Typically forms anhedral grains in rocks. In veins, it may form large crystals flattened on {010}.

Composition: Deviates very little from its formula.

Distinguishing Properties: Low birefringence and negative relief characterize gypsum and distinguish it from anhydrite. Gypsum has strong dispersion (r>b).

Occurrence: The main occurrence is as a sedimentary mineral in evaporite deposits, where it is commonly associated with anhydrite. It also occurs with limestones and shales and as a cement in some sandstones.

G
L

LAWSONITE $CaAl_2(OH)_2[Si_2O_7]H_2O$ Biaxial + Orthorhombic

(MM 50, 51)

Refractive Index Relief	Birefringence Interf. Color	2V	Extinction Elongation
x 1.66	0.02		parallel length
y 1.67		80	length slow
z 1.68	2nd order blue		

Color Pleochroism	Cleavage	Cleavage Angle	Hardness
mostly colorless			
x pale blue	{100} perfect	{100}^{010} 90	6
y pale green	{010} perfect		
z colorless			

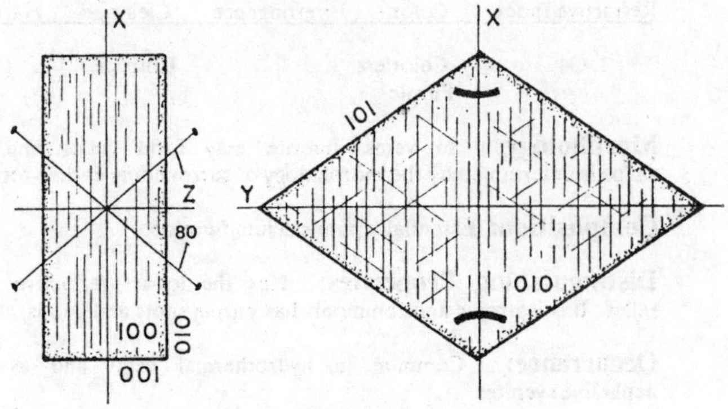

Morphology: Euhedral crystals with rhombic or rectangular outline. Single and lamellar twins on {101} are common.

Composition: Lawsonite has a composition that can be thought of as hydrated anorthite. Indeed, heating of lawsonite causes it to convert to anorthite.

Distinguishing Properties: Lawsonite resembles clinozoisite but lacks the anomalous interference colors. Lawsonite has better cleavage and crystal shape than clinozoisite. Lawsonite has very strong dispersion (r>b).

Occurrence: Forms at low temperatures and high pressures. A common constituent of glaucophane schists.

LEUCITE $K[AlSi_2O_6]$ Pseudo Isotropic (+) Tetragonal

(IM 68, 69)

Refractive Index	Color	Birefringence	Cleavage	Hardness
1.51	colorless	dark gray	none	6

Morphology: Euhedral crystals with octagonal outline. Above 600°C leucite is cubic and grows as trapezohedral crystals. On cooling it becomes tetragonal, and fine lamellar twinning develops in numerous sectors within a crystal.

Composition: Shows little deviation from its formula but may contain a small amount of Na substituting for K.

Distinguishing Properties: Faint lamellar twinning in an almost isotropic mineral and the crystal outline.

Occurrence: Restricted almost entirely to highly potassic volcanic rocks where it forms phenocrysts. It is commonly associated with aegirine. It is also found in shallow potassic intrusions, but here it commonly inverts or reacts with the magma to form intergrowths of nepheline and orthoclase. These intergrowths retain the morphology of the original leucite crystals and are therefore referred to as pseudoleucite.

MELILITE $(Ca,Na)_2[(Mg,Al,Si)_3O_7]$

Uniaxial + -   Tetragonal

(IM 71, 72; MM 40)

Refractive Index Relief	Birefringence Interf. Color	Extinction Elongation
Mg Al	Mg Al	
e 1.64-1.66	0.008-0.011	parallel
o 1.63-1.67	anomalous 1st order	length slow

Color Pleochroism	Cleavage	Cleavage Angle	Hardness
colorless			
e	{001} poor		5
o			

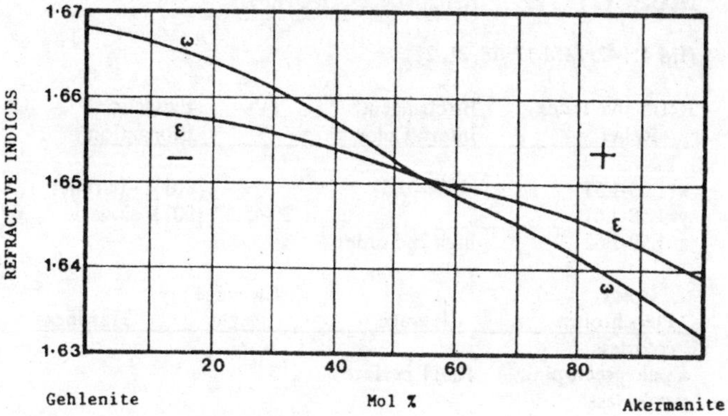

Morphology: Rectangular crystals flattened parallel to (001). Also as anhedral grains.

Composition: The melilites have three principal end-members with complete solid solution between them all.

Akermanite	Ca_2	Mg	Si_2O_7
Gehlenite	Ca_2	Al	$AlSiO_7$
Soda melilite	CaNa	Al	Si_2O_7

Melilites with compositions between akermanite and gehlenite are common in high-temperature metamorphosed impure dolomites. Melilites in igneous rocks contain a considerable proportion of soda melilite.

M

Distinguishing Properties: All melilites have low birefringence (maximum, 1st order gray) and are uniaxial. Gehlenite-rich varieties are negative, whereas akermanite-rich ones are positive. The change from negative to positive occurs approximately half way through this series. Intermediate members are therefore almost isotropic and show anomalous blue and brown interference colors. Melilite differs from nepheline, with which it is commonly associated, by having much higher refractive indices. Apatite has hexagonal cross sections and does not exhibit anomalous interference colors. Melilite occurs only in rocks with low silica activity. For example, it is never found with feldspar, but it is with nepheline; it does not occur with sphene, but it does with perovskite.

Occurrence: Melilites between gehlenite and akermanite are common constituents of high temperature metamorphosed impure dolomite, where they may be associated with monticellite, diopside, or forsterite. Melilite also occurs in highly silica-undersaturated alkaline igneous rocks, where it usually contains a large amount of soda mellilite.

MICA

Micas are present in a wide compositional range of igneous and metamorphic rocks, and they are stable over a considerable range of temperatures and pressures. Their perfect basal cleavage, which is a consequence of the sheets of linked $(Si,Al)O_4$ tetrahedra, makes them one of the easiest groups of minerals to identify in both hand specimen and in thin section.

Micas can be divided into two subgroups, di-octahedral and tri-octahedral, on the basis of the number of octahedral atoms between the tetrahedral layers. The di-octahedral micas have trivalent octahedral ions (mainly Al,Fe''') and the tri-octahedral micas have divalent ions (mainly Mg,Fe''), thus maintaining charge balance. Micas are essentially potassium-bearing minerals, but in the di-octahedral ones some sodium may substitute for potassium. A miscibility gap separates sodium-bearing muscovite from the essentially sodium-bearing di-octahedral mica paragonite. In the tri-octahedral micas complete solid solution exists between the Mg and Fe end members of the series. In both the di- and tri-octahedral micas K can be replaced by Ca to form the rarer brittle micas margarite and clintonite respectively.

Micas and amphiboles form the two important groups of hydrous rock-forming minerals. Micas are stable over a wider range of conditions than are amphiboles, forming in such diverse environments as sedimentary rocks and upper mantle peridotites. Although the di- and tri-octahedral micas both contain aluminum, the di-octahedral ones have an excess over the amount of alkalis. Di-octahedral micas are thus more common in alumina-rich rocks. Muscovite, for example, is the main mineral in most low- to intermediate-grade meta-pelites. Muscovite also occurs in many granites where Al is in excess over alkalis. The genesis of these granites likely involves the incorporation of a significant amount of pelitic rocks. Biotite is also common in granites and many metamorphic rocks, but it does not require an excess of alumina for its formation. Biotite is a common constituent of alkaline mafic igneous rocks, and in some may even form phenocrysts. The Mg-rich variety, phlogopite, commonly occurs in marbles and is a principal constituent of the rock type kimberlite, in which diamonds are found.

MUSCOVITE $KAl_2[AlSi_3O_{10}](OH,F)_2$

Biaxial - Monoclinic

(IM 41, 42; MM 17, 18, 21, 22)

Refractive Index Relief	Birefringence Interf. Color	2V	Extinction Elongation
x 1.55-1.57 y 1.58-1.61 z 1.59-1.62	0.04-0.05 high 2nd order	 30-45	{001} ~parallel {001} slow

Color Pleochroism	Cleavage	Cleavage Angle	Hardness
colorless x pale green-pink y colorless z colorless	 {001} perfect		 3

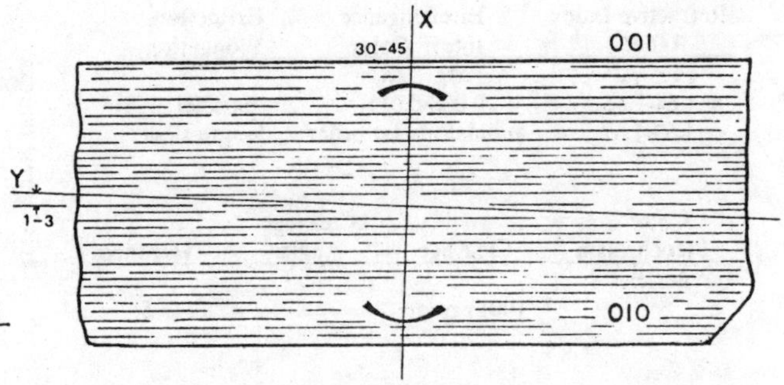

Morphology: Forms tabular crystals parallel {001}; these may be six-sided or irregular in outline. In pegmatites muscovite may form six-sided prisms that are elongated parallel to the c axis.

Composition: Muscovites show considerable variation in composition, but this is not easily determined from optics alone. Na substituting for K produces a limited range of solid solution towards paragonite, the Na analogue of muscovite. Some muscovites contain higher proportions of silica in the tetrahedral sites and are referred to as high-silica muscovite or phengite. In these, charge balance is maintained by substitution of Mg or Fe" for Al in the octahedral sites--these, consequently, tend to be pinkish or greenish in hand specimen.

Distinguishing Properties: Easily distinguished by its high birefringence, essentially parallel extinction along its only cleavage direction, {001}. The cleavage direction is length slow. Muscovite exhibits well the phenomenon known as birdseye extinction, which is characteristic of all the micas. When rotated to extinction the grain appears mottled because of the failure of small spots to go to extinction. It resembles the wood grain in birdseye maple. Because sections cut parallel to {001} have such low interference colors (1st order creamy white) and completely lack any signs of cleavage, they appear to be of a completely different mineral from the muscovite grains that are oriented so as to reveal the cleavage. It is these basal sections that show a very striking bisectrix figure with isochromatic rings. The 2V of normal muscovite is about 45 but with increasing silica content the 2V decreases to as low as 15 in phengite. This is still larger than the 2V of most phlogopite, which is usually considerably less than 15.

Occurrence: A common constituent of pelitic schists formed at low- to medium-grades of regional metamorphism. It is also an important constituent of pegmatites and aluminous granites where it is commonly associated with biotite.

PARAGONITE $NaAl_2[AlSi_3O_{10}](OH,F)_2$

Biaxial - Monoclinic

Refractive Index Relief	Birefringence Interf. Color	2V	Extinction Elongation
x 1.56-1.58 y 1.59-1.61 z 1.60-1.61	0.03 2nd order yellow	 0-40	{001} ~parallel {001} slow

Color Pleochroism	Cleavage	Cleavage Angle	Hardness
colorless x y z	 {001} perfect		 2.5

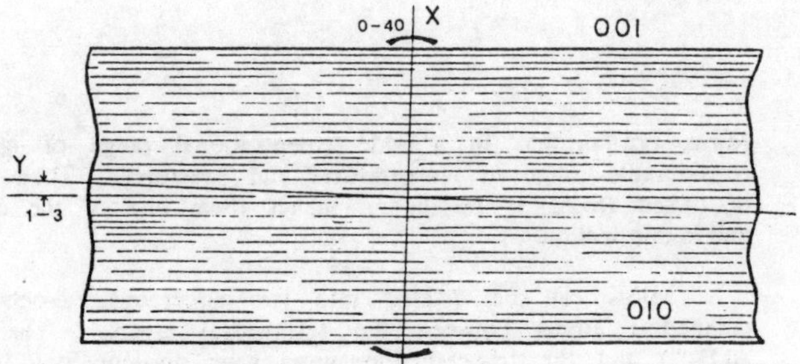

Morphology: Forms tabular crystals flattened on {001}.

Composition: At high temperatures paragonite can dissolve up to 20% of the muscovite component. At low temperatures, however, it approaches closely the formula above.

Distinguishing Properties: Without chemical analysis paragonite cannot be distinguished for certain from muscovite using only optical properties. The 2V of most paragonite is smaller than that of muscovite, but high-silica muscovite also has a small 2V.

Occurrence: Paragonite occurs with muscovite in greenschist metamorphic rocks. It may be more abundant than presently recognized because of the difficulty of distinguishing it optically from muscovite. With increasing metamorphic grade, paragonite becomes more soluble in muscovite. Because muscovite is typically the more abundant of the two minerals, paragonite disappears as a separate phase by dissolving in the muscovite.

BIOTITE $K(Mg,Fe)_3[AlSi_3O_{10}](OH,F)_2$

Biaxial - Monoclinic

(IM 39-43; MM 17-20)

Refractive Index Relief	Birefringence Interf. Color	2V	Extinction Elongation
x 1.56-1.63 y 1.60-1.70 z 1.60-1.70	0.04-0.07 2nd-3rd order	0-25	{001} 0 - 9 {001} slow

Color Pleochroism	Cleavage	Cleavage Angle	Hardness
red,brown,green x light .. y z dark ..	{001} perfect		2-3

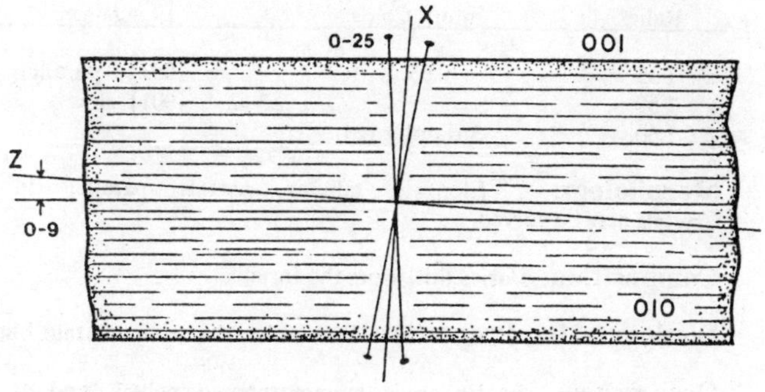

Morphology: Commonly occurs as well-formed tabular crystals in pelitic metamorphic rocks and in granites. In many gabbroic rocks it crystallizes late and is anhedral. In some alkaline mafic rocks, especially lamprophyres, it may crystallize early and form tabular phenocrysts.

Composition: Biotites range widely in composition with the main substitutions being Mg for Fe and Al for Si, with the latter also requiring substitution of trivalent ions (Fe''', Al) into the octahedral sites. Substitution of Ti turns biotite a deep red, and Fe''' turns it green. The lithium-rich tri-octahedral mica is lepidolite.

Distinguishing Properties: Most easily distinguished by color and pleochroism, prominent {001} cleavage, and high birefringence and extinction nearly parallel to {001}. The only common mineral resembling biotite is hornblende; biotite differs from it in having only one cleavage direction, nearly parallel extinction, and a stippled appearance in the extinction position known as birdseye texture, which is characteristic of all the micas.

Occurrence: One of the most common of the rock-forming minerals. It occurs in all but the very lowest and highest metamorphic grade pelitic rocks. In many granites it may be the major ferromagnesian mineral. It is a major constituent of rocks of intermediate composition, such as granodiorites, diorites, monzonites, and some nepheline syenites. In most mafic igneous rocks it is a late crystallizing accessory phase, which commonly forms as a rim around magnetite and ilmenite grains. In some lamprophyric dikes (minette, kersantite, alnoite) biotite forms phenocrysts that commonly have a rounded core surrounded by a strongly compositionally-zoned rim. Highly magnesian biotite occurs in kimberlite (see phlogopite).

PHLOGOPITE $K(Mg,Fe)_3[AlSi_3O_{10}](OH,F)_2$

Biaxial - Monoclinic

Refractive Index Relief	Birefringence Interf. Color	2V	Extinction Elongation
x 1.53-1.59 y 1.56-1.64 z 1.56-1.64	0.03-0.05 2nd-3rd order	0-15	{001} 0 - 5 {001} slow

Color Pleochroism	Cleavage	Cleavage Angle	Hardness
x colorless y pale brown z pale brown	{001} perfect		2-3

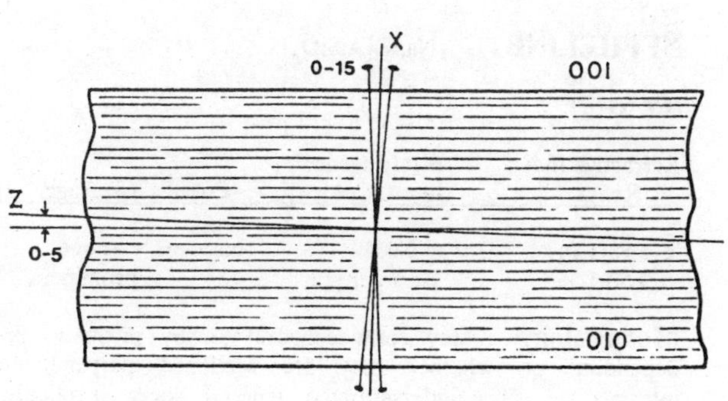

Morphology: Thick tabular crystals are common in marbles.

Composition: Magnesium end member of the tri-octahedral micas; usually restricted to those with Mg:Fe > 2.

Distinguishing Properties: Distinguished from muscovite by smaller 2V and from chlorites by higher birefringence.

Occurrence: The main occurrence is in contact metamorphosed marbles and regionally metamorphosed impure dolomitic limestones; it is thus commonly associated with calcite and calc-silicates. Phlogopite occurs in some potassic ultramafic rocks, especially those containing leucite. It is also a major constituent of the groundmass of kimberlite breccias.

PYROPHYLLITE $Al_4[Si_8O_{20}](OH)_4$ Biaxial - Monoclinic

Refractive Index Relief	Birefringence Interf. Color	2V	Extinction Elongation	Color	Cleavage	Hardness
x 1.55	0.05		{001} ~parallel	colorless	{001} perfect	1-2
y 1.59		53-62	{001} slow			
z 1.60	3rd order red					

Morphology: Elongate tabular crystals parallel to {010}, radiating clusters, and fine-grained aggregates. Crystals may be curved.

Composition: Varies little from the formula.

Distinguishing Properties: Resembles muscovite and talc, but has a larger 2V.

Occurrence: In low-grade metamorphosed pelites, and as a hydrothermal alteration of feldspar, where it is commonly associated with quartz.

MONTICELLITE $CaMg[SiO_4]$ Biaxial - Orthorhombic

(IM 15)

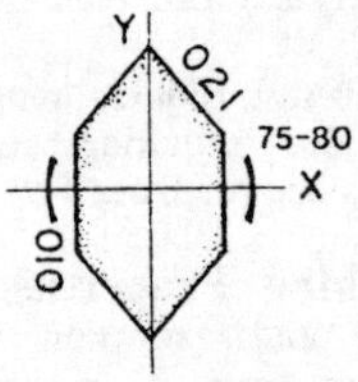

Refractive Index Relief	Birefringence Interf. Color	2V	Extinction Elongation	Cleavage	Hardness
x 1.64-1.65	0.01-0.02		parallel	none	5.5
y 1.65-1.66		75-80			
z 1.65-1.67	1st order red				

Morphology: Colorless crystals with the same shape as those of olivine. In undersaturated alkaline igneous rocks and carbonatites it can form phenocrysts. More commonly it forms anhedral grains with curving fractures lined with dark alteration products. In high-temperature contact metamorphosed limestone it forms granular grains.

Composition: May contain a small amount of Fe substituting for Mg.

Distinguishing Properties: Under plane light monticellite and olivine appear identical. Under crossed polars, however, monticellite has a much lower birefringence. Optic axis figures of monticellite do not exhibit the numerous isochromatic rings that those of olivine do.

Occurrence: In igneous rocks it is restricted to those that are extremely undersaturated in silica. It does not occur with feldspars but does with melilites. It is a common constituent of carbonatites and high-temperature contact metamorphosed impure dolomitic limestones.

NEPHELINE $(Na,K)[AlSiO_4]$ Uniaxial - Hexagonal

(IM 63-67)

Refractive Index Relief	Birefringence Interf. Color	Color	Cleavage	Hardness
e 1.53-1.54	0.005	colorless	none	6
o 1.53-1.55	1st order gray		prominent	

Morphology: May form euhedral stubby prismatic phenocrysts which exhibit hexagonal or rectangular cross sections depending on whether they are sectioned perpendicular or parallel to the c axis respectively. In mafic and intermediate silica-undersaturated igneous rocks, nepheline crystallizes late and thus takes on the shape imposed by surrounding crystals.

Composition: Most nepheline contains approximately one K for every three Na ions in the structure. With increasing temperature nepheline can dissolve more of the K end-member, kalsilite. Kalsilite itself is a rare mineral but does form exsolution lamellae in nepheline, and it does form separate crystals in some rare potassic volcanic rocks and in contact metamorphosed limestone. Kalsilite contains very little Na.

Distinguishing Properties: Low birefringence, low relief, and uniaxial negative figure are characteristic. Because of low birefringence, it goes into extinction easily and when totally black commonly exhibits bright birefringent flecks of muscovite which are a common alteration product of nepheline.

Occurrence: A common mineral in silica undersaturated rocks. A major constituent of nepheline syenite and occurs in smaller amounts in nepheline gabbro (essexite) and its volcanic equivalent basanite.

OPAQUES

A number of common accessory minerals and important ore minerals are opaque. Their microscopic identification can be made with certainty only in polished sections under reflected light. Grain morphology and associated nonopaque alteration products may serve to distinguish some of these minerals in transmitted light. Polished thin sections are useful because they permit examination of both opaque and nonopaque minerals in reflected and transmitted light. The following table lists the most useful optical properties for identifying the common opaque minerals. Most of these properties are seen under reflected light, but some may be evident with transmitted light.

Under reflected light, two properties, the color of the mineral and whether it is isotropic or anisotropic, greatly narrow the search for the identity of an unknown mineral. The colors of opaque minerals under reflected light are mostly gray or cream. With few exceptions, the gray ones are oxides and the cream ones are sulfides. This distinction can sometimes be made with an ordinary thin section by illuminating it obliquely from above. Both groups contain cubic and noncubic minerals, which can be distinguished by their reflection anisotropism. Remember that with reflected light, the polarizers are uncrossed by a few degrees to permit sufficient light through the microscope to make the outlines of grains visible. Isotropism or anisotropism is judged on the basis of changes in color or brightness rather than on total extinction. Relative hardness of grains can be judged by the relief on the polished surface. If grains are large enough their hardness can be checked by trying to scratch them with a needle.

Mineral	Formula	Color in plane polarized light	Reflectivity	Optical Character	Hardness Mohs	Hardness Talmage	Other Distinguishing Properties
Magnetite	$Fe^{2+}OFe^{3+}{}_2O_3$	Gray, brownish tint	>ilmenite <<hematite	Isotropic	6	F	Commonly forms octahedra. May contain lamellae of ilmenite, hematite, or hercynite.
Chromite	$Fe^{2+}OCr_2O_3$	Dark gray, brownish	<<magnetite <<ilmenite	Isotropic	8	G	Deep brown in transmitted light. Forms octahedra.
Hematite	$Fe^{3+}{}_2O_3$	Bright grayish white	>>magnetite >>ilmenite	Strongly Anisotropic	6	F-G	Hexagonal plates. Red in transmitted light on thin edge.
Ilmenite	$Fe^{2+}OTiO_2$	Grayish brown with pinkish or violet tint	<magnetite <<hematite	Strongly Anisotropic	5-6	G	Hexagonal plates. Deep brown in transmitted light on very thin edges. May contain exsolution lamellae of hematite. Alters to leucoxene, a white Ti-oxide.
Graphite	C	Gray, brownish tint	=magnetite >ilmenite	Strongly Anisotropic	1-2	A	Hexagonal plates. Polishes with difficulty. Reflection pleochroism is very strong.
Sphalerite	ZnS	Gray	<magnetite =ilmenite	Isotropic	4	C	Deep reddish brown color in transmitted light. {110} cleavage prominent.
Galena	PbS	White	extremely bright	Isotropic	2.6	B	Polishes easily but often has scratches. Extremely soft. Cubic cleavage always present.
Pyrite	FeS_2	Yellowish white	extremely bright	Isotropic	6	F-	Polishes with difficulty. Cube and pyritohedron are common forms. Alters to limonite.
Pyrrhotite	$Fe_{1-x}S$ ($x = 0 - 0.2$)	Cream with pinkish brown tint	<pyrite	Strongly Anisotropic	4	D-	Polishes easily. Shows yellow, green, and blue colors under crossed polars. Alters readily to limonite.
Arseno-pyrite	FeAsS	White with cream or pink tint	=pyrite >pyrrhotite	Strongly Anisotropic	6	F	Yellow, green, and blue are evident under crossed polars. Lamellar twinning very common.
Chalco-pyrite	$CuFeS_2$	Brassy yellow	<pyrite >pyrrhotite	Weakly Anisotropic	4	C	Polishes easily. Lamellar twins common.

OLIVINE $(Mg,Fe)_2[SiO_4]$

Biaxial + -　　　　Orthorhombic

(IM 5-15; MM 35,36)

Refractive Index Relief	Birefringence Interf. Color	$2V_z$	Extinction Elongation
Mg Fe	Mg Fe	Mg Fe	
x 1.63-1.83	0.03-0.05		parallel cleavage
y 1.65-1.87		82-134	
z 1.67-1.88	2-3 order		

Color Pleochroism	Cleavage Cleavage	Angle	Hardness
colorless-yellow			
x	{010} poor	{010}^{100} 90	7-6.5
y	{100} poor		
z			

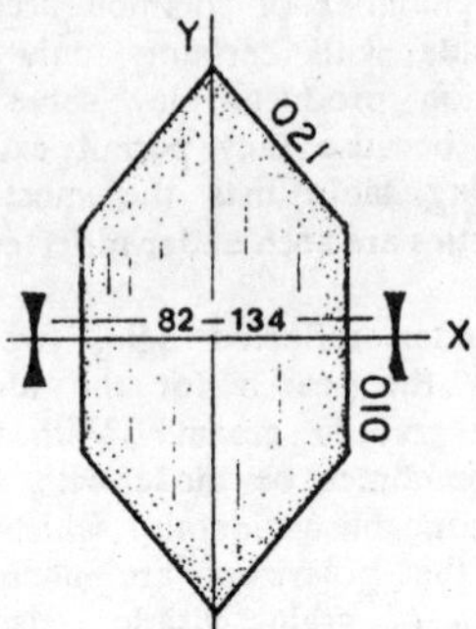

Morphology: Commonly forms phenocrysts in igneous rocks where it typically has the prismatic form illustrated in the figure. May show a faint rectangular cleavage, but curving fractures lined with alteration products, such as serpentine, magnetite, talc, and carbonate, are more common. In the Archean ultramafic lava komatite, olivine forms extremely long dendritic crystals. In many plutonic igneous rocks and in metamorphic rocks, olivine is anhedral.

Composition: Simple compositions between the two end-members Forsterite (Mg) and Fayalite(Fe). As shown in the plot below, optical properties of olivine vary linearly with substitution of Fe for Mg. The composition of olivine can therefore be determined easily under the microscope. These compositions are normally quoted as the mole percent Fo. Fe-rich varieties are more yellow. Olivine commonly contains significant amounts of Ni.

Distinguishing Properties: Easily identified by high birefringence, lack of marked cleavage, curving fractures filled with dark alteration products, and large 2V. In volcanic rocks, olivine may enclose patches of trapped melt.

Occurrence: One of the most common minerals in mafic and ultramafic igneous rocks. It is also thought to be a major constituent of the Earth's mantle. Olivine is also a product of the metamorphism of dolomitic rocks.

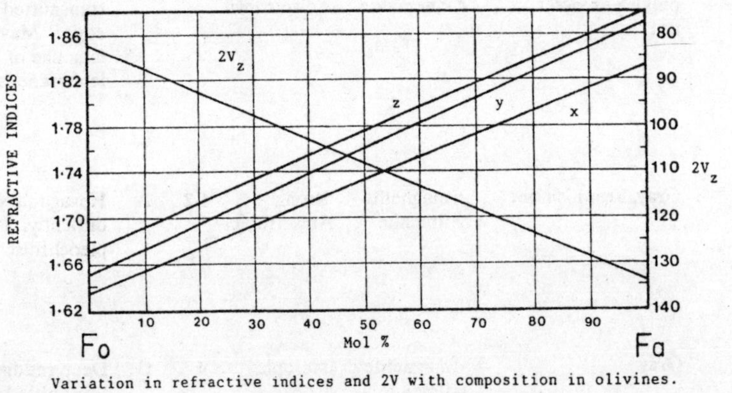

Variation in refractive indices and 2V with composition in olivines.

PEROVSKITE $CaTiO_3$

Pseudo-isotropic +　　　　Monoclinic

Refractive Index	Color	Birefringence	Cleavage	Hardness
2.30-2.38	yellow-brown	weak	poor	5.5

Morphology: Small cubic crystals that stand out because of their very high relief.

Composition: A large number of elements can substitute into perovskites, in particular the rare earths for Ca, and Nb and Ta for Ti.

Distinguishing Properties: The very high relief, color, crystal form, and weak birefringence of large crystals (with faint lamellar twinning) are typical of perovskite.

Occurrence: Restricted to rocks with very low activities of silica. It is not found in rocks that contain feldspar. It is associated with melilite, nepheline, monticellite, and calcite. It occurs in both igneous and metamorphic rocks.

PYROXENE

Pyroxenes fall into three compositional types, the Ca-poor, the Ca-rich, and the Na-rich. At high temperature solid solution between these three is extensive, but at low temperature a large miscibility gap separates Ca-poor and Ca-rich varieties, especially in pyroxenes with a high Mg/Fe ratio. Consequently, many Ca-poor and Ca-rich pyroxenes that cool slowly from high temperatures develop exsolution textures, which can provide information about the initial conditions of formation. The three main types of pyroxene are distinguished easily with optical measurements, and even the composition within a particular series can be determined approximately from microscope measurements. The Ca-poor and Ca-rich varieties commonly occur together in subalkaline igneous rocks and their high-grade metamorphic equivalents. The Ca-poor pyroxenes do not occur in alkaline (and Si-undersaturated) igneous rocks. Instead, these rocks contain pyroxenes that are commonly zoned from a Ca-rich core to a Na-rich rim. The sodic pyroxene jadeite forms only at very high pressures. The compositional range of the non-sodic pyroxenes is conveniently represented in the lower half of a ternary diagram with Wollastonite at the top and Enstatite and Ferrosilite at the base. The Ca-rich pyroxenes, Diopside and Hedenbergite, fall half way along the sides of the triangle, and with Enstatite and Ferrosilite form the end-members of a compositional representation commonly referred to as the pyroxene quadrilateral.

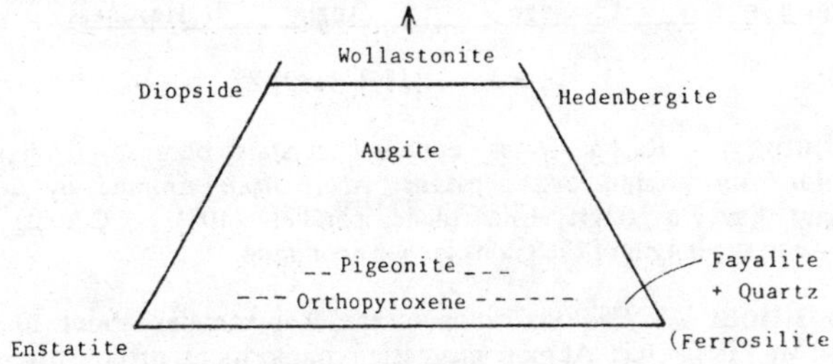

The general formulae and some important optical properties of the pyroxenes are as follows:

		Symmetry	Sign	2V	z^c	Color
Orthorhombic Pyroxene	$(Mg,Fe)_2[SiO_3]_2$	Orthorhombic	-(+)	50-90	0	colorless
Pigeonite	$(Mg,Fe,Ca)_2[SiO_3]_2$	Monoclinic	+	0-30	40	colorless
					x^c	
Clinopyroxene - Augite	$Ca(Mg,Fe)[SiO_3]_2$	Monoclinic	+	~50	45	colorless
Aegirine (Acmite)	$NaFe^{3+}[SiO_3]_2$	Monoclinic	-	65	<10	green
Jadeite	$NaAl[SiO_3]_2$	Monoclinic	+	70	33-40	colorless

ORTHOPYROXENE $(Mg,Fe)_2[SiO_3]_2$

(IM 16-24)

Biaxial - (+) Orthorhombic

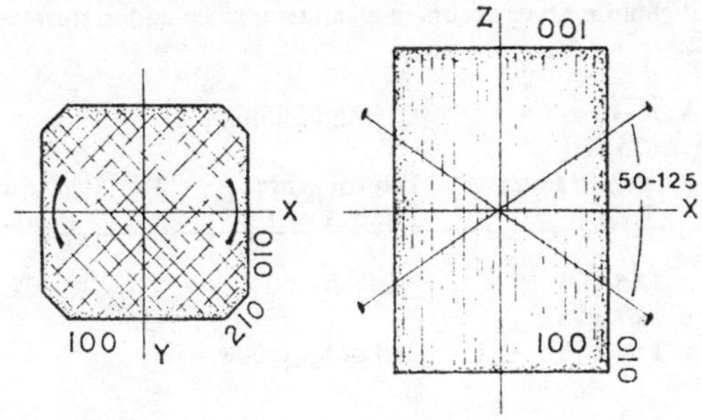

Refractive Index Relief	Birefringence Interf. Color	2V	Extinction Elongation
x 1.65-1.77 y 1.65-1.77 z 1.66-1.79	0.01-0.02 1st order yellow	50-90	parallel c axis c axis slow

Color Pleochroism	Cleavage	Cleavage Angle	Hardness
colorless or pale x pale pink y pale yellow z pale green	{210} good	{210}^{2$\bar{1}$0} 88	5-6

Morphology: Commonly forms euhedral prismatic crystals, elongated parallel to c, with eight-sided cross sections. Sections cut parallel to {010} may have a streaky appearance under crossed polars produced by thin lamellae of slightly varying birefringence parallel to {100}. Thin exsolution lamellae of augite commonly parallel {100}.

Composition: Vary from the pure Mg end-member, Enstatite, to compositions generally no more Fe-rich than about 90% of the Fe-rich end-member, Ferrosilite. Under crustal pressures the Fe-rich orthopyroxenes are unstable and fayalite + quartz forms instead. At pressures greater than 1.15 GPa, orthoferrosillite is stable. The orthopyroxenes are given specific names for particular ranges of mole percentage of Mg/(Mg+Fe); enstatite is from 100 to 88, bronzite from 88 to 70, hypersthene from 70 to 50, ferro-hypersthene from 50 to 30, eulite from 30 to 12, and ferrosilite from 12 to 0. When possible, however, it is preferable to give the composition of an orthopyroxene in terms of its percentage of enstatite, which is expressed as $En_{\%}$. In mafic igneous rocks, orthopyroxenes are typically in the bronzite to hypersthene range. In metamorphic rocks they have a wider range of compositions.

Distinguishing Properties: Distinguished by low birefringence, especially in sections cut parallel {010}, where the section may exhibit streaky lamellae with white and gray interference colors. Sections cut parallel to the c axis have parallel extinction and are lenght slow.

Occurrence: Bronzite and hypersthene are common in tholeiitic gabbros and diabases and calc-alkali andesites. Orthopyroxenes in igneous rocks are rarely more Fe-rich than hypersthene; instead pigeonite or inverted pigeonite is the Ca-poor pyroxene formed at these Fe-rich compositions. In high-grade metamorphic rocks, orthopyroxenes can have compositions throughout the entire range, being limited at the Fe-rich end by the reaction to form fayalite + quartz.

PYROXENE

PIGEONITE $(Mg,Fe,Ca)_2[SiO_3]_2$ Biaxial + Monoclinic
(IM 30,31)

Refractive Index Relief	Birefringence Interf. Color	2V	Extinction Elongation
x 1.68-1.72	0.02-0.03		z^c 37-44
y 1.68-1.72		0-30	length fast
z 1.70-1.75	2nd order green		

Color Pleochroism	Cleavage	Cleavage Angle	Hardness
colorless	{110} good	{110}^{1$\bar{1}$0} 87	6

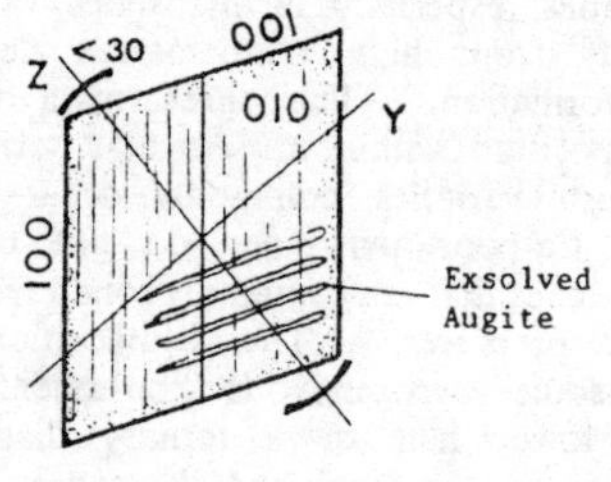

Morphology: Rarely forms euhedral crystals because it crystallizes late between earlier-formed crystals; grows as a reaction rim around orthopyroxene, or is itself rimmed by augite. Grains tend to be elongated parallel to c and commonly have a single twin plane parallel {100}. Curving fractures (often filled with dark alteration products) transverse to the length of the grains are very common.

Composition: A Ca-poor clinopyroxene that contains about 10 mole % of the wollastonite component. Most pigeonite has $Mg/(Mg+Fe) < 0.7$. At more magnesian compositions, orthopyroxene is the common Ca-poor pyroxene.

Distinguishing Properties: Characterized by small 2V, commonly appearing almost uniaxial. The only other pyroxene with small 2V is Ti-rich augite, but this is distinctly brownish, pinkish, or violet and has anomalous low interference colors. Pigeonite is clearer than augite, with which it is almost always associated, but pigeonite alters more readily to dark products along cleavages and curving fractures. Its birefringence is slightly less than that of coexisting augite, but the difference is so small that it serves to distinguish the minerals only where they are in optical continuity. Pigeonite is commonly rimmed by augite, which grows in optical continuity with the pigeonite.

Occurrence: The common Ca-poor pyroxene in subalkaline rocks that have $Mg/(Mg+Fe)$ less than ~0.7. Stable only at high temperatures, and preserved only in rapidly-cooled igneous rocks, such as lava flows, dikes, and small intrusive bodies. On slow cooling it first exsolves lamellae of augite nearly parallel to {001} but then inverts to orthopyroxene. Orthopyroxene contains less Ca than pigeonite and, thus, the inversion involves further exsolution of augite which is added to the already-formed lamellae in the pigeonite. Pigeonite consequently inverts to orthopyroxene containing thick augite lamellae nearly parallel to the original {001} of pigeonite. Such orthopyroxene grains are referred to as inverted pigeonite (IM 32, 33). In some inverted pigeonite augite lamellae are quite irregular or bleb-like. Further cooling of the orthopyroxene results in thin lamellae of augite exsolving nearly parallel to {100}. Pigeonite never occurs in alkaline or silica undersaturated rocks. It forms in some high-temperature metamorphic rocks.

AUGITE $Ca(Mg,Fe)[SiO_3]_2$ Biaxial + Monoclinic
(IM 25, 27)

Refractive Index Relief	Birefringence Interf. Color	2V	Extinction Elongation
x 1.66-1.73	0.02-0.03		z^c ~ 45
y 1.67-1.74		~50	
z 1.69-1.76	2nd order yellow		

Color Pleochroism	Cleavage	Cleavage Angle	Hardness
almost colorless			
x pale brown-green	{110} good	{110}^{1$\bar{1}$0} 87	6
y pale brown			
z pale green			

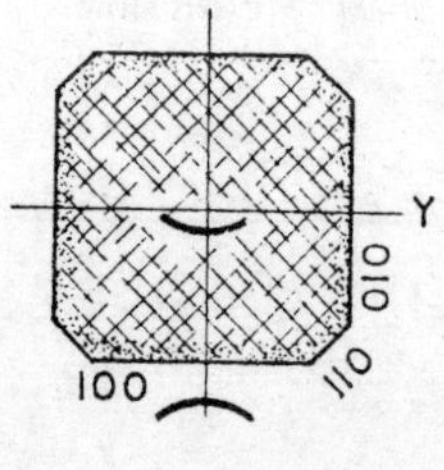

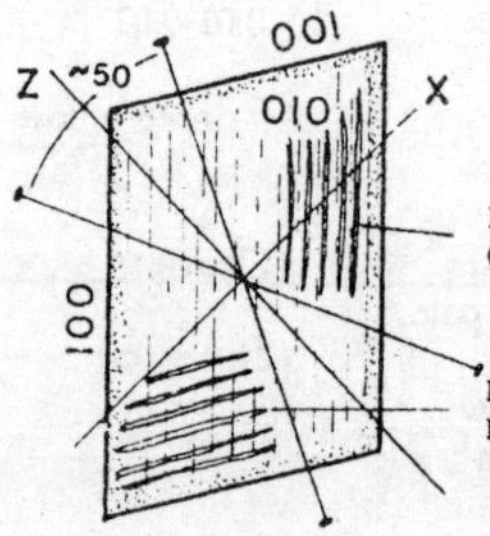

Morphology: In mafic alkaline igneous rocks and andesites augite commonly forms stubby euhedral prisms. In more siliceous rocks the prisms are elongated. In tholeiitic rocks augite tends to be anhedral. In marble, it commonly forms globular grains with a vitreous surface.

Composition: Complete solid solution between the Mg end-member, Diopside, and the Fe end-member, Hedenbergite. Intermediate compositions in tholeiitic rocks typically contain less Ca than do pyroxenes on the diopside-hedenbergite join, and they coexist with a Ca-poor pyroxene. Augite coexisting with pigeonite will have exsolution lamellae of pigeonite nearly parallel to {001}, whereas that coexisting with orthopyroxene will contain orthopyroxene lamellae nearly parallel to {100}. In alkaline igneous rocks, augites have compositions very near the diopside-hedenbergite join but may also contain Na and Ti.

Distinguishing Properties: Distinguished from pigeonite by larger 2V, slightly more color, and being less altered.

Occurrence: One of the major constituents of mafic alkaline and subalkaline igneous rocks. In the alkaline rocks it may contain considerable amounts of Na or Ti (see next page). In subalkaline rocks it is associated with a Ca-poor pyroxene (orthopyroxene or pigeonite) or olivine or both. Augite formed in metamorphosed impure dolomitic limestones is typically rich in the diopside component. Augite in mafic igneous rocks metamorphosed in the granulite facies has the same general range of composition as that in the igneous rocks but with slightly higher Ca contents.

TITANAUGITE $(Ca,Mg,Fe^{2+},Ti,Al)_2[(Si,Al)O_3]_2$

Biaxial + Monoclinic

(IM 26, 27)

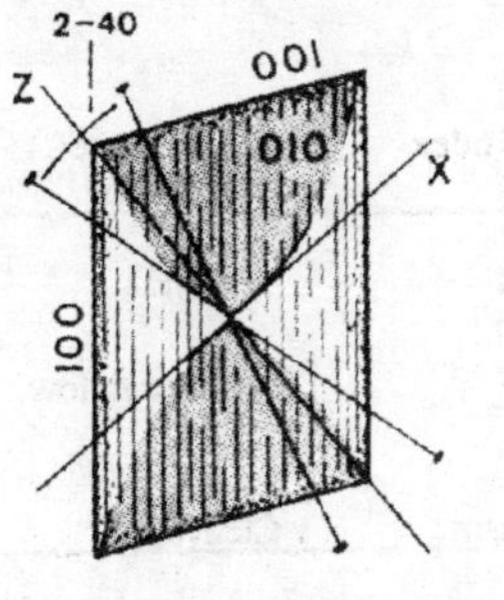

Refractive Index Relief	Birefringence Interf. Color	2V	Extinction Elongation
x 1.70-1.74	0.02-0.03		z^c 30-40
y 1.72-1.74		2-40	length slow
z 1.73-1.76	2nd order yellow		

Color Pleochroism	Cleavage	Cleavage Angle	Hardness
purplish brown			
x pale brown	{110} good	{110}^{1$\bar{1}$0} 87	6
y brown-violet			
z purple			

Morphology: Commonly forms euhedral stubby prisms that exhibit sector zoning (hourglass structure) which is made evident by slight differences in birefringence between sectors. Crystals also commonly show fine oscillatory zoning.

Composition: Ti probably enters the structure as $CaTiAl_2O_6$. As a result, titanaugites have low silica contents. Some Ti may even enter the tetrahedral site and replace Si.

Distinguishing Properties: Resembles augite, except that its color is much darker and the 2V is much smaller. Pigeonite also has a small 2V but is almost colorless. Titanaugite has strong dispersion (r>b).

Occurrence: Restricted to silica-undersaturated rocks and is therefore commonly associated with feldspathoids, melilites, perovskite, and other undersaturated minerals. It never occurs with Ca-poor pyroxene.

AEGIRINE (Acmite) $NaFe^{3+}[SiO_3]_2$

Biaxial - Monoclinic

(IM 28)

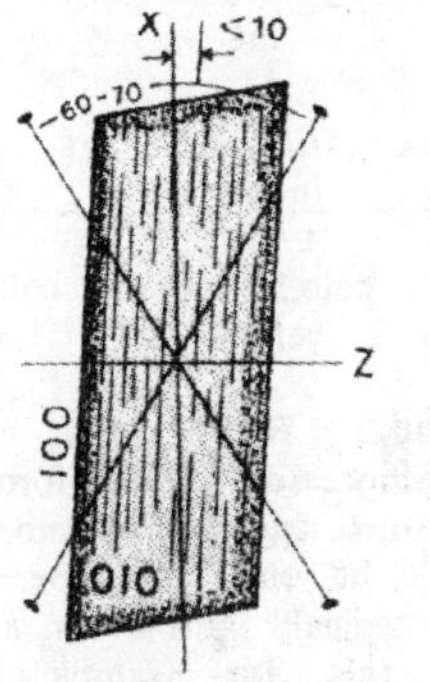

Refractive Index Relief	Birefringence Interf. Color	2V	Extinction Elongation
x 1.75-1.78	0.04-0.05		x^c < 10
y 1.78-1.82		60-70	length fast
z 1.80-1.84	3rd order yellow		

Color Pleochroism	Cleavage	Cleavage Angle	Hardness
grass green			
x dark green	{110} good	{110}^{1$\bar{1}$0} 87	6
y light green			
z yellowish green			

Q

Morphology: Typically crystallizes as long prisms but may also form rims on stubby augite crystals.

Composition: There is complete solid solution from augite to aegirine, with the color changing from clear to green with increasing substitution of NaFe''' for Ca(Mg,Fe''). Crystals are commonly zoned from augite at the core out through aegirine-augite to aegirine on the rim.

Distinguishing Properties: Distinguished by bright transparent grass green color; green amphiboles are less transparent and appear muddy by comparison. Cleavage angles distinguish aegirine from green amphiboles. The small extinction angle distinguishes aegirine from other clinopyroxenes. Intermediate compositions between aegirine and augite have intermediate optical properties.

Occurrence: Occurs in Na-rich igneous rocks. It may be associated with nepheline and sodalite in nepheline syenites, but it also occurs with quartz in alkali syenites and granites.

JADEITE $NaAl[Si_2O_6]$

PYROXENE

Biaxial + Monoclinic

(MM 52)

Refractive Index Relief	Birefringence Interf. Color	2V	Extinction Elongation
x 1.64-1.66	0.01		c^z 33-40
y 1.65-1.66		70	length slow
z 1.65-1.67	1st order yellow		

Color Pleochroism	Cleavage	Cleavage Angle	Hardness
colorless x to y pale z green	{110} good	{110}^{1$\bar{1}$0} 87	6

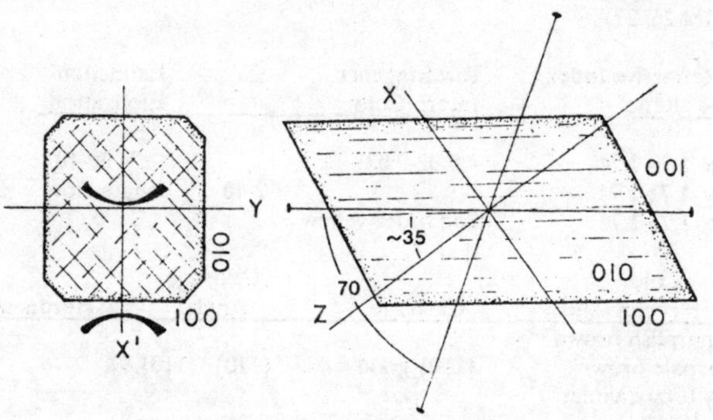

Morphology: Columnar to fibrous aggregates, rarely with euhedral form.

Composition: May show solid solution towards diopside; in eclogites this solution is extensive, forming the pyroxene omphacite.

Distinguishing Properties: Has pyroxene cleavage but differs from the other pyroxenes in having lower refractive indices and much lower birefringence. Orthopyroxene also has low birefringence but has parallel extinction. Dispersion in jadeite is moderate (r>b).

Occurrence: Stable only at high pressures; breaks down at low pressure to albite plus nepheline. Jadeite is therefore restricted to metamorphic rocks formed at high pressures, such as the glaucophane schists and eclogites. Omphacite occurs in mantle-derived eclogite nodules in kimberlite.

QUARTZ SiO_2

SILICA

Uniaxial + Trigonal

(IM 59-61)

Refrac. Index Relief	Birefring. Interf.	Color	Cleavage	Hardness
e 1.55 o 1.54	0.01 pale 1st order yellow	colorless	none	7

Morphology: Rarely forms euhedral crystals except in some veins. Two polymorphs, the low-temperature alpha and the high-temperature beta forms, have different crystal morphologies. All quartz at room temperature is the low-temperature trigonal polymorph, regardless of the temperature of formation, because the inversion, which occurs at about 573°C at one atmosphere, is spontaneous. However, quartz formed in the stability field of the hexagonal beta polymorph typically grows as a dipyramid, and this morphology is preserved after inversion to the low-temperature polymorph; this, for example, is the common form of quartz phenocrysts in rhyolites. Quartz growing in the low-temperature field is more likely to form elongate prisms, as in typical vein quartz. Many quartz grains in igneous, metamorphic, and sedimentary rocks consist of numerous domains with slightly different extinction position. This results from strain and is known as undulatory extinction. It is so common in quartz that it provides a diagnostic optical property. The variety of quartz known as Chalcedony consists of fine fibres, which exhibit a spherulitic texture under crossed polars. This form commonly fills cavities and fractures in igneous rocks and replaces calcareous fossils in sedimentary rocks. Chalcedony commonly exhibits alternating clear and brown layers parallel to the surface on which it is deposited.

Composition: Shows almost no compositional deviation from its formula.

Distinguishing Properties: Easily identified through low relief and complete lack of alteration products under plane light, and through grey to pale yellow interference colors that usually exhibit undulatory extinction under crossed polars. It is uniaxial positive whereas nepheline and scapolite are uniaxial negative. Most nepheline also exhibits alteration, and most scapolite has higher birefringence. The only clouding found in quartz is due to small fluid inclusions which typically consist of a gas bubble suspended in liquid. If the bubble is small it may move around in the fluid as a result of thermal vibrations.

Occurrence: An extremely common mineral in continental crustal rocks. A major constituent of granites and of detrital sedimentary rocks and their metmorphosed equivalents. It is the most common vein-forming mineral.

Several other polymorphs of SiO_2 exist, but they are much less common than quartz. At atmospheric pressure and temperatures above 867°C, tridymite replaces quartz as the stable polymorph, and above 1470°C cristobalite becomes the stable form. The inversion from quartz to tridymite is raised significantly by pressure, whereas the inversion from tridymite to cristobalite is lowered. As a result, the stability field of tridymite is eliminated above a pressure of 0.14 GPa. Cristobalite is limited to pressures of less than 0.6 GPa. Above 2.5 GPa the polymorph coesite is stable, and above about 7 GPa a still higher-pressure polymorph, stishovite, is stable. These high-pressure polymorphs have been found around meteorite impact craters, and coesite occurs in some eclogites. Although the transition from high- to low-quartz is spontaneous, the other transitions are sluggish, and as a result the other polymorphs may persist in rocks to low temperatures and pressures. Cristobalite and tridymite, however, can both form metastably outside their stability fields, and thus their presence in a rock is not positive evidence of formation at high temperatures. By contrast, the high-pressure polymorphs form only at high pressure.

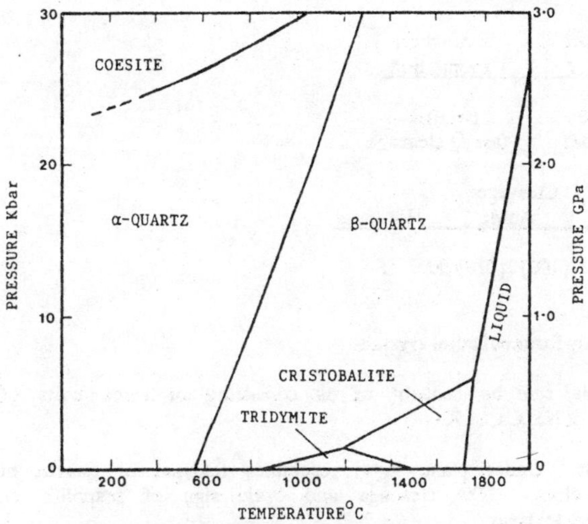

Stability fields of the silica polymorphs.
(After Tuttle and Bowen, 1958; Cohen and Klement, 1967; Morse, 1980; Bohlen and Boettcher, 1982)

TRIDYMITE Orthorhombic Biaxial + 2V = 35 R.I. ~1.47 Birefringence extremely low

(IM 62)

Typically forms elongated wedge-shaped crystals with a twin boundary along the length of the wedge. It has very low relief. Occurs in cavities in siliceous volcanic rocks. Some quartz in granophyres has the elongate wedge-like morphology of tridymite and it likely inverted from this polymorph. Tridymite is a common constituent of refractory bricks.

S

CRISTOBALITE Tetragonal Esentially Isotropic R.I. ~1.48 Birefringence extremely low

Forms small cube-like crystals and occurs in cavities in siliceous volcanic rocks with tridymite and sanidine.

COESITE Monoclinic Biaxial + 2V = 64 R.I. ~1.595 Birefringence very low (0.005)

Distinguished by high relief, very low birefringence, and being colorless. Occurs as extremely small grains in quartz-bearing rocks that have been exposed to high shock pressures by meteorite impact; this variety is normally identifiable only by x-ray diffraction. Also found as equant grains with grossular-pyrope garnet, omphacite, and sanidine in mantle nodules in kimberlite. Also occurs as inclusions in diamonds. Rare metasedimentary blueschist rocks can contain coesite as inclusions in pyrope-rich garnet. In all occurrences, coesite is rimmed and veined by polyhedral quartz that has formed by inversion from the high-pressure polymorph. Because of the difference in refractive indices between these two minerals the inversion rims and veins stand out in striking relief. Inversion from coesite to quartz involves considerable expansion which causes radiating tension fractures to form in the surrounding minerals, especially garnet.

RUTILE TiO_2 Uniaxial + Tetragonal

Refractive Index Relief	Birefringence Interf. Color	Extinction Elongation	Color Pleochroism	Cleavage	Hardness
			reddish brown		
e 2.90	0.29	parallel	brownish	{110} good	6
o 2.61		length slow	yellowish		

Common accessory mineral in igneous, metamorphic, and sedimentry rocks. In igneous and metamorphic rocks it forms very small needles that are mostly enclosed in other minerals, particularly quartz. Its birefringence is so high that in normal thickness sections little color is seen under crossed polars especially if masked by the mineral's usual reddish brown color. Most crystals, however, are so small that their thickness is only a fraction of that of the section, and these may exhibit second or third order interference colors. In sedimentary rocks it forms detrital grains.

SCAPOLITE $Na_4[Al(Al,Si)Si_2O_8]_3(Cl,CO_3,SO_4,OH)$ Uniaxial - Tetragonal

(MM 37)

Refractive Index Relief	Birefringence Interf. Color	Extinction Elongation
e 1.54-1.57	0.005-0.036	parallel
o 1.55-1.60	1st-2nd order	fast // cleavage

Color Pleochroism	Cleavage	Cleavage Angle	Hardness
colorless			
e -	{100} good	{100}^{010} 90	6
o -	{110} poor		

Morphology: Does not typically form euhedral crystals.

Composition: The formula can be thought of as consisting of three units of feldspar and one of a chloride, carbonate, sulphate,or hydroxide of Na, Ca, or K.

Distinguishing Properties: Under plane light, resembles feldspar or quartz, but under crossed polars has bright blue to green interference colors. The cleavage and optic sign of scapolite distinguish it from quartz, and the uniaxial figure distinguishes it from feldspar.

Occurrence: Forms instead of feldspar in some metamorphic rocks where there is a high concentration of carbonate or sulphate. Common in marbles and some amphibolites, and particularly common in contact metamorphic skarns.

SERPENTINE $Mg_3[Si_2O_5](OH)_4$ Biaxial - Monoclinic

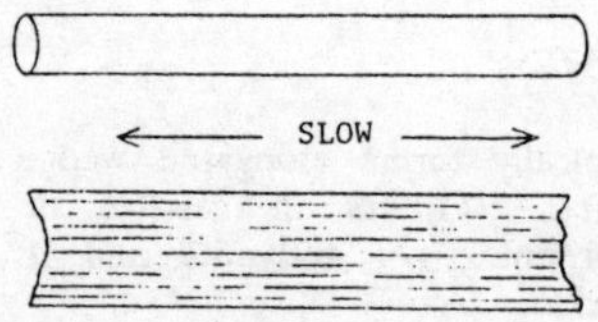

(IM 12; MM 35, 36)

Refractive Index Relief	Birefringence Interf. Color	2V	Extinction Elongation
x 1.53-1.57	< 0.01		parallel
y 1.57	max. is	~50	length slow
z 1.54-1.57	1st order yellow		

Color Pleochroism	Cleavage	Cleavage Angle	Hardness
colorless-pale green			
colorless-pale green	{001} perfect		3

Morphology: Forms aggregates of platy, fibrous, or amorphous-looking material. Several different forms are commonly juxtaposed. It replaces olivine and pyroxene, and may appear to have the euhedral shape of these minerals. Under crossed polars, however, such pseudomorphs can be seen to consist of a multitude of small grains with such varied orientations that no clear extinction occurs. The variety of serpentine known as chrysotile forms veins of fibrous material that is length slow. This is the chief source of asbestos.

Composition: Differs little from its formula.

Distinguishing Properties: Chlorite resembles serpentine, but chlorite has higher relief, is mostly length fast, and typically is more colored and pleochroic. Serpentine does not have anomalous interference colors.

Occurrence: Forms by alteration of olivine and pyroxene, and as the low-grade metamorphic product of rocks rich in these minerals such as peridotites and pyroxenites. It typically pseudomorphically replaces these minerals. The original mineral can be identified either by the external morphology of the pseudomorph or the texture of the serpentine itself. Serpentine that replaces olivine commonly retains the pattern of curving fractures that are present in most olivine grains, whereas serpentine formed in pyroxene has a more rectangular pattern.

SODALITE $Na_3[AlSiO_4]_3NaCl$ Isotropic Cubic

Refractive Index	Color	Birefringence	Cleavage	Hardness
1.48-1.49	colorless	none	{110} poor	6

Morphology: Forms six-sided dodecahedral crystals, but also forms anhedral grains interstitial to other minerals.

Composition: The formula can be thought of as consisting of 3 nepheline + 1 NaCl. The Cl can be replaced by SO_4 to form the mineral Nosean. If the Na is replaced by Ca the mineral Hauyne is formed. Complete solid solution exists between these end-members.

Distinguishing Properties: Difficult to distinguish from analcite in thin section, but in hand specimen sodalite is commonly blue. Nosean and hauyne commonly contain abundant dark inclusions that may be concentrated into sectors, planes, cores, or rims. Nosean and hauyne are often blue in thin section but sodalite is colorless.

Occurrence: Occurs in Na-rich silica undersaturated felsic rocks such as nepheline syenite and sodalite syenite. Nosean and hauyne are common in phonolites.

SPHENE (Titanite) $CaTi[SiO_4](O,OH,F)$ Biaxial + Monoclinic

(IM 77, 78)

Refractive Index Relief	Birefringence Interf. Color	2V	Extinction Elongation
x 1.84-1.95	0.10-0.19		cannot easily
y 1.87-2.03	extremnely high	20-40	be determined
z 1.94-2.11	white		

Color Pleochroism	Cleavage	Hardness
brownish yellow		
x almost colorless	{110} distinct	5
y pale yellow		
z brownish		

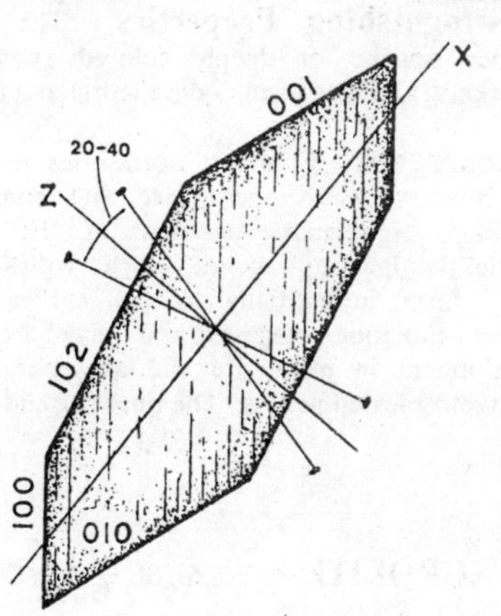

Morphology: Forms distinctive wedge- or elongated rhomb-shaped crystals.

Composition: Substitution of Ca by rare earths and of Ti by Fe, Mg, Al, Nb, and Ta in sphene is common.

Distinguishing Properties: The wedge or elongated rhomb shape of the crystals is very characteristic. With the exception of zircon, sphene is the only brownish colored mineral that appears the same under plane or crossed polarized light. The birefringence is so high that the white interference color contributes nothing to the brown color of the mineral. Zircon can be distinguished by its crystal shape (prisms with square cross sections) and by its uniaxial character. Sphene has such high dispersion (r>b) that grains commonly do not extinguish completely. A white alteration product, leucoxene, may be visible on the surface of sphene when illuminated obliquely from above.

Occurrence: A common accessory in many igneous and metamorphic rocks, and also occurs as detrital grains in sediments. It can be a major constituent in metamorphic rocks with high Ca and Ti contents, such as meta basalts and pyroxenites and calc-silicates.

SPINEL $(Mg,Fe^{2+})(Fe^{3+},Cr,Ti,Al)_2O_4$ Isotropic Cubic

(IM 80)

The spinel series consists of a large group of minerals which form common accessories in rocks and some ore deposits. All are cubic, and most are either opaque or deeply colored, the end-member spinel being the exception–it is pale colored or colorless. Opaque spinels must be examined under reflected light (see opaque minerals), studied with x-rays, or analyzed under the electon microprobe to determine their precise character. The nonopaque spinels, on the other hand, can be easily identified under the microscope by their color, high refractive index (>1.72), and isotropic character. Some of the more important members of the series and their colors in thin section are as follows:

Spinel	$MgAl_2O_4$	Colorless to pale green
Hercynite	$Fe^{2+}Al_2O_4$	Deep green
Magnesioferrite	$MgFe^{3+}{}_2O_4$	Dark brown to opaque
Magnetite	$Fe^{2+}Fe^{3+}{}_2O_4$	Opaque
Ulvospinel	$Fe^{2+}Fe^{2+}TiO_4$	Opaque
Chromite	$(Mg,Fe^{2+})Cr_2O_4$	Brown

Morphology Most form small, equant, octahedral grains. Exsolution of one spinel from another may produce lamellae or granules.

Composition At high temperatures there is extensive solid solution in the spinel series, and with the exception of magnetite and spinel itself pure end-members are not found in nature. With falling temperature exsolution may take place. For example, magnetite commonly contains exsolution lamellae of hercynite. It may also contain lamellae of ulvospinel, but these can be seen only under reflected light. Oxidation of magnetite-ulvospinel solid solutions results in the formation of lamellae of ilmenite in the magnetite, but these are not exsolution lamellae because there is no solid solution between magnetite and ilmenite.

Distinguishing Properties The occurrence of spinels as small high refractive index octahedral grains that are either opaque or deeply colored serves to distinguish them from other minerals. The octahedra give 4- and 6-sided sections. Garnets form dodecahedral and trapezohedral crystals which have 6- and 8-sided sections.

Occurrence Common accessories in igneous, metamorphic, and sedimentary rocks. Magnetite is the most common, and in most rocks is the phase that contains most of the ferric iron. In alkaline and calc-alkaline igneous rocks, which contain significant amounts of Fe''', magnetite tends to crystallize early, forming octahedral phenocrysts. In tholeiitic igneous rocks, which typically contain much smaller concentrations of Fe''', magnetite crystallizes late and may form interstitially to the earlier minerals. Chromite, which has a very high melting point, crystallizes early in some tholeiitic magmas and may accumulate by settling to form ore bodies. Hercynite occurs as a solid solution component in magnetite and as separate grains in some mafic and ultramafic igneous rocks. It also forms in high grade metamorphosed pelites. The pure Mg end-member, spinel, occurs in contact metamorphosed limestone.

STAUROLITE $Fe_2Al_9Si_{3.75}O_{22}(OH)_2$ Biaxial + Monoclinic

(MM 24, 25)

Refractive Index Relief	Birefringence Interf. Color	2V	Extinction Elongation
x 1.74-1.75 y 1.74-1.75 z 1.75-1.76	0.013 1st order yellow	 80-90	parallel length slow

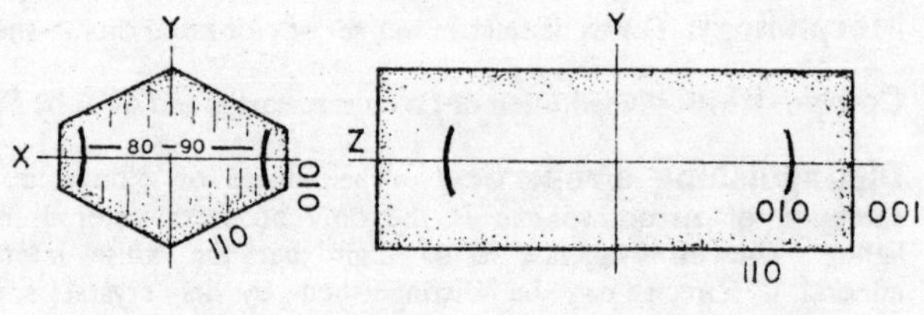

Color Pleochroism	Cleavage	Cleavage Angle	Hardness
yellow x colorless y pale yellow z yellow	{010} poor		7.5

Morphology: Normally forms euhedral prismatic crystals with six-sided cross sections. The crystals are mostly larger than those of other minerals in the rock (porphyroblasts). Despite its euhedral habit, staurolite commonly contains many inclusions of other minerals, especially quartz.

Composition: The composition of staurolite is close to that of the formula. A small amount of Mg may substitute for Fe, and some Fe''' can substitute for Al. Small amounts of Zn substitute for Fe.

Distinguishing Properties: Easily identified by its yellow color and pleochroism.

Occurrence: Strictly a metamorphic mineral in pelitic rocks in the lower to middle amphibolite facies.

TALC $Mg_3[Si_4O_{10}](OH)_2$

Biaxial - Monoclinic

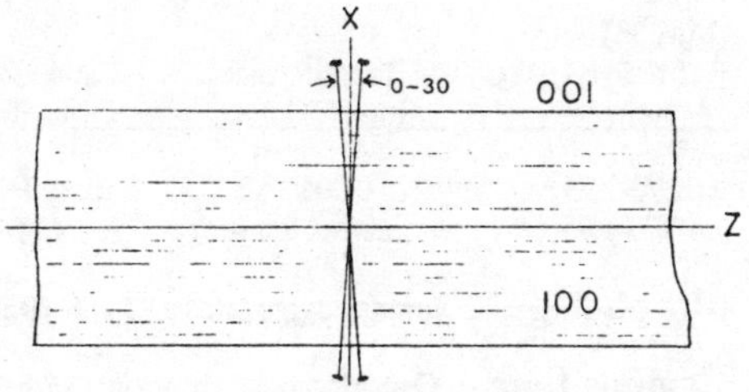

	Refractive Index / Relief	Birefringence / Interf. Color	2V	Extinction / Elongation
x	1.54-1.55	0.05		parallel {001}
y	1.59		0-30	{001} slow
z	1.59-1.60	3rd order red		

	Color / Pleochroism	Cleavage	Cleavage Angle	Hardness
	colorless			
x		{001} perfect		1
y				
z				

Morphology: Forms platy and fibrous aggregates.

Composition: Differs little from its formula.

Distinguishing Properties: Characterized by high birefringence and micaceous or fibrous appearance. Difficult to distinguish from muscovite, but talc has a smaller 2V than most muscovite. The association of other magnesian minerals is the best indication that the phase may be talc; x-ray or electron probe analysis may be necessary for confirmation.

Occurrence: Formed by hydrothermal alteration of ultramafic rocks and by the metamorphism of impure dolomitic limestone. It commonly forms in shear zones in these rocks.

TITANITE (see Sphene)

TOURMALINE $Na(Mg,Fe,Al)_3Al_6[Si_6O_{18}](BO_3)_3(OH,F)_4$

Uniaxial - Trigonal

(IM 79)

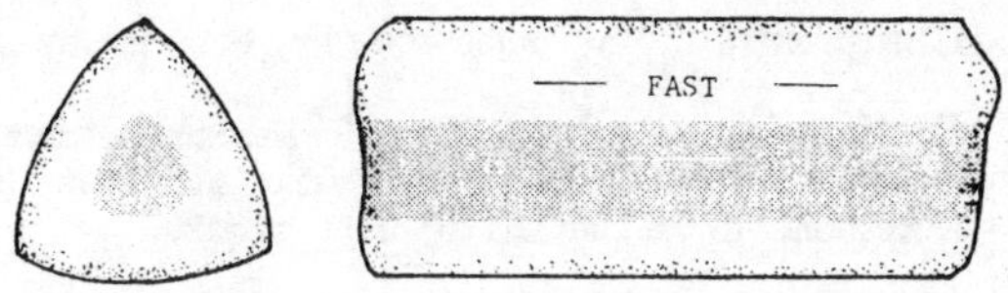

	Refractive Index / Relief	Birefringence / Interf. Color	Extinction / Elongation
e	1.61-1.65	0.02-0.035	parallel
o	1.63-1.67	2nd order	length fast

	Color / Pleochroism	Cleavage	Hardness
	olive-blue-yellow		
e	pale	none	7
o	dark		

Morphology: Typically forms prismatic crystals with curved three-sided cross sections. Crystals are commonly zoned, which is particularly evident in cross sections.

Composition: Highly variable in composition. Crystals that do not exhibit zoning are unusual. The main variation is in the Fe and Mg contents. The Fe-rich variety, schorl, is the common black tourmaline which appears various shades of green in thin section. Paler varieties are richer in the Mg end-member, draivite. Other substitutions involve Fe''', Al, Cr, and Li.

Distinguishing Properties: Strong pleochroism which, in contrast to that in the other common pleochroic rock-forming minerals (such as biotite and hornblende), shows its maximum absorption perpendicular to the length of the crystal. Tourmaline does not exhibit cleavage, whereas biotite and hornblende do. Tourmaline also is uniaxial, but in dark varieties the interference figures may be masked by the color of the mineral.

Occurrence: The most common boron-containing mineral. A commøn accessory in granites, and in many metamorphosed sedimentary rocks. It occurs as detrital grains in sedimentary rocks. Tourmaline is abundant in pegmatites.

VESUVIANITE (Idocrase) $Ca_{19}(Mg,Fe,Al)_{13}Si_{18}(O,OH,F)_{76}$ — Uniaxial - — Tetragonal

(MM 38)

Refractive Index Relief	Birefringence Interf. Color	Extinction Elongation	Color Pleochroism	Cleavage	Hardness
e 1.70-1.75 o 1.70-1.75	0.004 anomalous 1st order	parallel length fast	colorless to pale yellow	{110} poor	6-7

Morphology: Columnar aggregates and anhedral grains.

Composition: Considerable variation, with Na, K, and Mn substituting for Ca, and Al, Ti, Zn and Mn substituting for Mg and Fe.

Distinguishing Properties: Much vesuvianite exhibits anomalous blue and brown interference colors. It is distinguished from negative varieties of melilite by its higher refractive index. Clinozoisite resembles vesuvianite but is biaxial. Unfortunately some varieties of vesuvianite are biaxial. These varieties are also difficult to distinguish from grossular garnet, which can also be biaxial.

Occurrence: The principal occurrence is in contact metamorphosed limestones and skarn deposits. It also occurs in regionally metamorphosed limestones and in serpentinites.

WOLLASTONITE $CaSiO_3$ — Biaxial - — Triclinic

(MM 39)

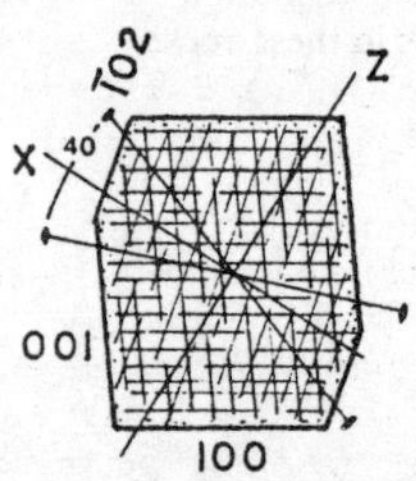

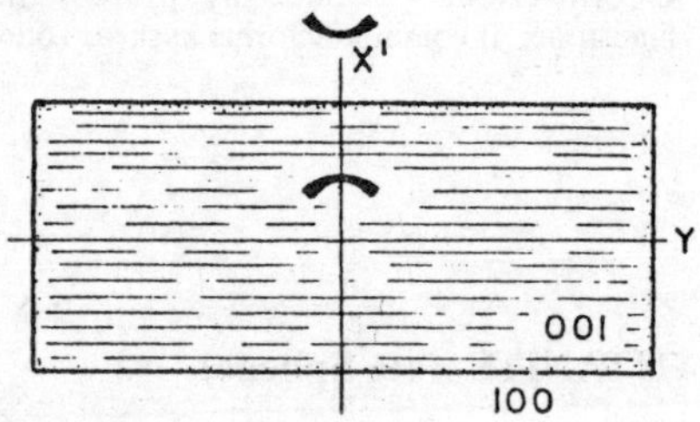

Refractive Index Relief	Birefringence Interf. Color	2V	Extinction Elongation
x 1.62-1.64 y 1.63-1.65 z 1.63-1.65	0.013 1st order orange	~40	parallel length slow and fast

Color Pleochroism	Cleavage Cleavage	Angle	Hardness
colorless x - y - z -	{100} perfect {001} good {$\bar{1}$02} good	{100}^{001} 84 {100}^{$\bar{1}$02} 70	5

Morphology: Columnar and fibrous aggregates with twinning common parallel to {100}.

Composition: May contain some iron but is mostly near the formula in composition.

Distinguishing Properties: Resembles tremolite, but wollastonite has extinction parallel to the length of the crystals. Wollastonite crystals also are both length slow and length fast, because the y axis of the indicatrix corresponds to the long axis (b) of the crystals.

Occurrence: A metamorphic mineral in high grade contact metamorphosed impure limestone.

ZIRCON $ZrSiO_4$ — Uniaxial + — Tetragonal

(IM 73, 74)

Refractive Index Relief	Birefringence Interf. Color	Extinction Elongation	Color	Cleavage	Hardness
e 1.97-2.01 o 1.92-1.96	0.04-0.06 4th order pink	parallel length slow	pale brown	none	7.5

Morphology: Most forms small stubby prismatic grains with square cross sections. Where enclosed in ferromagnesian minerals, they are commonly surrounded by a pleochroic halo.

Composition: Most zircon contains small amounts of U and Th, which, on undergoing radioactive decay, destroy the structure of the surrounding zircon (metamict). The radiocative decay also develops pleochroic haloes in juxtaposed ferromagnesian minerals. Most zircon also contains some Hf.

Distinguishing Properties: High relief, high birefringence, and crystal shape.

Occurrence Common as an accessory in granitic igneous rocks, as detrital grains in sedimentary rocks, and as an accessory in metamorphic rocks formed from these.

ANALCITE $Na[AlSi_2O_6].H_2O$ Isotropic Cubic

(IM 70)

Refractive Index	Color	Birefringence	Cleavage	Hardness
1.48-1.49	colorless cloudy	anomalous	none	5.5

Morphology: Irregular masses in the groundmass of alkaline rocks and is thus bounded by faces of the surrounding crystals. In amygdules and veins it can form octagonal-outlined trapezohedral crystals.

Composition: Little variation in composition; some K and Ca substitute for Na.

Distinguishing Properties: Isotropic or only weakly birefringent. Fluorite has a much lower refractive index and thus has much stronger negative relief. Leucite shows anomalous birefringence and commonly has sectors with multiple twinning; most leucite occurs as well-formed primary crystals, whereas analcite is secondary. In thin section sodalite is similar to analcite but in hand specimen sodalite is commonly blue or pink, but it can be colorless like analcite.

Occurrence: Commonly occurs as a deuteric mineral in the groundmass of alkaline dyke and volcanic rocks. It may also form along with other zeolites and calcite as vein fillings.

ZEOLITES

(IM 70)

There are many zeolites, most of which have similar modes of occurrence; that is, as secondary minerals in mafic volcanic rocks, especially ones with alkaline affinities. They fill fractures and vesicles, or replace primary minerals. They are characterized by a framework structure with large holes that can contain H_2O -- 10 to 20 wt % H_2O. The water is easily driven out and may cause problems in the preparation of thin sections if samples are heated. In general, their compositions can be thought of as approximating hydrated feldspars and feldspathoids, minerals from which they are, in fact, derived by low temperature alteration or metamorphism in the zeolite facies.

Zeolites commonly grow in a variety of well-formed crystal shapes. They are colorless in thin section, have low negative relief, and most are biaxial with moderate to large optic angles. Many zeolites are not easily distinguished in the microscope on the flat stage without considerable effort and experience. Some of the most common, however, can be distinguished readily. These are listed below with their formulae and most characteristic properties.

Natrolite $Na_2[Al_2Si_3O_{10}].2H_2O$

Thomsonite $NaCa_2[(Al,Si)_5O_{10}]_2.6H_2O$

Laumontite $Ca[Al_2Si_4O_{12}].4H_2O$

Chabazite $Ca[Al_2Si_4O_{12}].6H_2O$

Heulandite $(Ca,Na_2)[Al_2Si_7O_{18}].6H_2O$

Stilbite $(Ca,Na_2,K_2)[Al_2Si_7O_{18}].7H_2O$

Mineral	2V	Sign	Morphology	Cleavage	Birefringence	Twinning	Extinction	Elongation
Natrolite	60	+	long prisms fibrous radiating	{110}^{110} 89	1st order orange		parallel	slow
Thomsonite	42-75	+	fibrous	{010} perfect	2nd order blue		parallel	fast and slow
Laumontite	26-47	-	prisms	{010}{110} good	1st order yellow	{100}	z^c 20-40	slow
Chabazite	0	+	cube-like rhombs	rhombohedral 85	1st order gray			
Heulandite	0-50	+	plates // {010}	{010} perfect	1st order white		parallel {010}	fast
Stilbite	30-50	-	sheafs	{010} good	1st order white	{001}	parallel {010}	fast and slow

REFERENCES

Bohlen, S. R., and Boettcher, A. L., 1982, The quartz=coesite transformation: A precise determination and the effects of other components: J. Geophys. Res., v. 87, p. 7073-7978.

Cohen, L. H., and Klement, W. Jr., 1967, High-low quartz inversion: Determination to 35 kilobars: J. Geophys. Res., v. 72, p. 4245-4251.

Holdaway, M. J., 1971, Stability of andalusite and the aluminum silicate phase diagram: Amer. J. Sci., v. 271, p. 97-131.

Morse, S. A., 1980, *Basalts and Phase Diagrams*, Springer Verlag, New York.

Tobi, A. C., and Kroll, H., 1975, Optical determination of the An-content of plagioclases twinned by Carlsbad-law: a revised chart: Amer. J. Sci., v. 275, p. 731-736.

Tuttle, O. F., and Bowen, N. L.,1958, Origin of granite in the light of experimental studies in the system $NaAlSi_3O_8$-$KAlSi_3O_8$-SiO_2-H_2O: Geol. Soc. Amer. Mem. 74, 153 p.

IGNEOUS ROCK-FORMING MINERALS

1. Properties such as color, relief, crystal form, cleavage, and relative crystal sizes are commonly most easily seen in the microscope under low magnification using plane polarized light. Visible in this section are two large crystals of hornblende (pleochroic green) with characteristic amphibole cleavage and small green aegerine crystals in a groundmass of alkali feldspar and nepheline (clear). Small crystals of apatite are included in the hornblende. Nepheline syenite, Mount Johnson, Quebec. Plane light, X8.

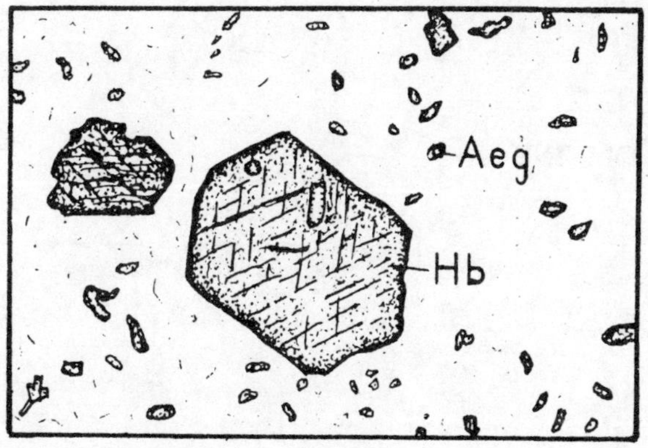

2. This metamorphic xenolith from a gabbro intrusion consists of two essential minerals, kalsilite (low relief) and diopside (high relief) and a minor opaque mineral. Kalsilite forms long bladed crystals, one of which extends from the lower left to the top center. Small rod-like inclusions of calcite mark the long axis of this crystal. Brome Mountain, Quebec. Plane light, X8.

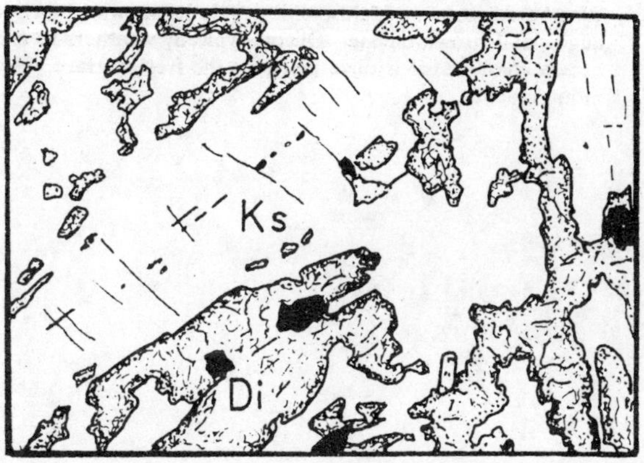

3. Same as in IM 2 but under crossed polars. Only the large kalsilite crystal is shown here to emphasize that what appears to be a single crystal under plane polarized light actually consists of a number of differently oriented (different shading) domains, formed when the kalsilite inverted on cooling from a high-temperature polymorph. This domain structure is visible only under crossed polars.

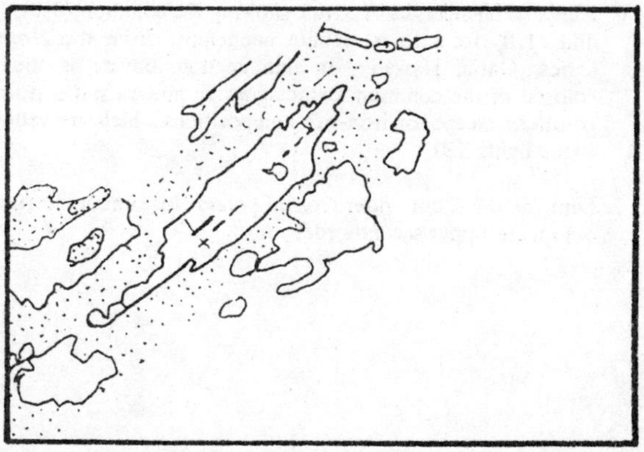

4. Knowledge of what mineral assemblages are possible and which are impossible can be of great benefit in doing petrography. For example, nepheline and quartz would not occur together in a rock because they would react to form albite. Consequently the assemblage nepheline plus albite or quartz plus albite are possible, but not nepheline plus quartz.

(Nepheline) (Quartz) (Albite)

$$NaAlSiO_4 + 2SiO_2 \rightarrow NaAlSi_3O_8$$

Compatable Assemblages:

Ne + Ab
Q + Ab

Incompatable Assemblage:

Ne + Q

OLIVINE

5. Weathered surface of feldspathic peridotite showing black titan-augite and brown olivine. Olivine typically weathers to a rusty brown color, but it is olive green on the fresh surface. Rougemont, Quebec.

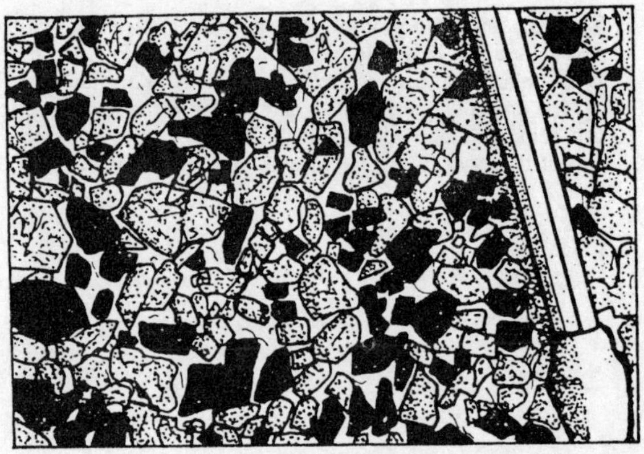

6. Euhedral phenocrysts of olivine showing the common (010), (021), and (110) faces in a melilite nephelinite from the Honolulu series, Oahu, Hawaii. In thin section, olivine is the least colored of the common ferromagnesian minerals; it is normally colorless except for iron-rich compositions which are yellowish. Plane light, X20.

7. Same as IM 6, but under crossed polars. Maximum interference colors are upper second order.

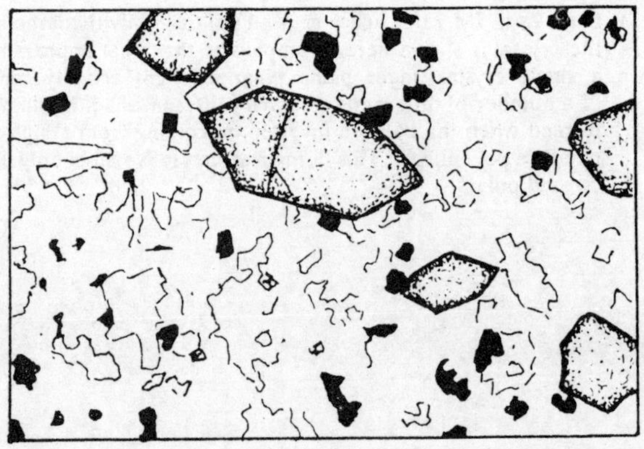

8. The shape of olivine crystals depends on their rate of growth. These olivine phenocrysts were already growing slowly in a basaltic magma before it was erupted on the flanks of Kilauea Volcano, Hawaii. The length to breadth ratio of these crystals is typical of slowly-grown olivine. Crossed polars, X20.

IGNEOUS ROCK-FORMING MINERALS

9. Olivine that grows rapidly commonly forms skeletal or dendritic crystals. These small olivine crystals grew rapidly in a basaltic pillow on the ocean floor at the Mid-Atlantic Ridge. Continued growth and filling in of skeletal crystals commonly lead to the trapping of melt inclusions in olivine crystals. Crossed polars, X80.

10. Olivine dendrites reach extreme length to breadth ratios in the Archean ultramafic lava known as komatiite where they form radiating blades extending down from the top of flows. The olivine here is intergrown with pyroxene and rapidly-quenched melt. Munro Township, Ontario. Crossed polars, X10.

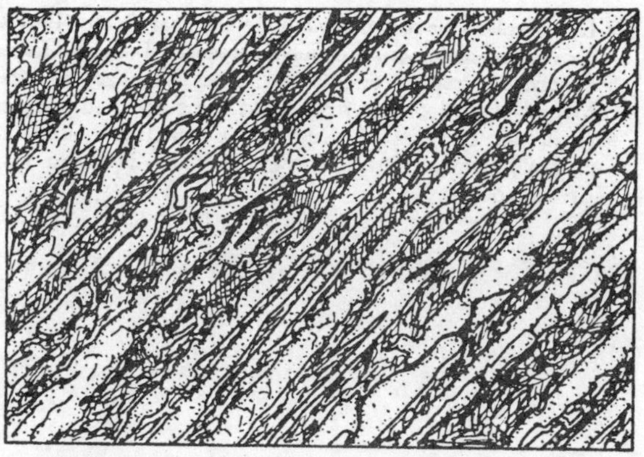

11. Cleavage in olivine is not common, but when present it forms in two directions intersecting at 90°. Extinction is parallel to these cleavages. In pyroxenes sections showing right angle cleavage have extinction at 45° to the cleavages. Cleavage in olivine phenocrysts is seen in this ankaramite from Maui, Hawaii. Crossed polars, X8.

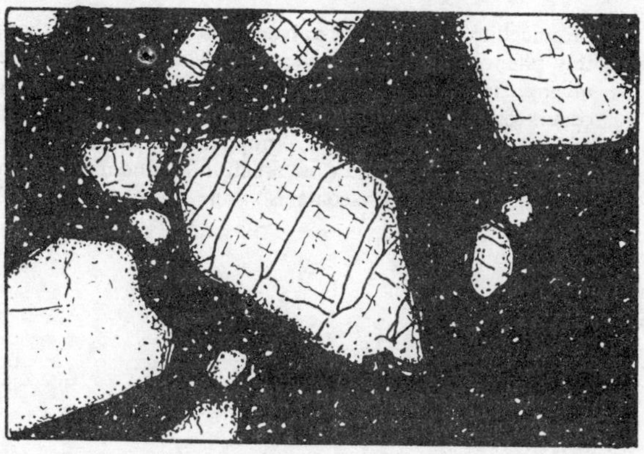

12. Talc pseudomorph after olivine that was first veined by serpentine. The birefringence of serpentine is very low, whereas that of talc is high. Grenville marble, Adirondacks, New York. Crossed polars, X8.

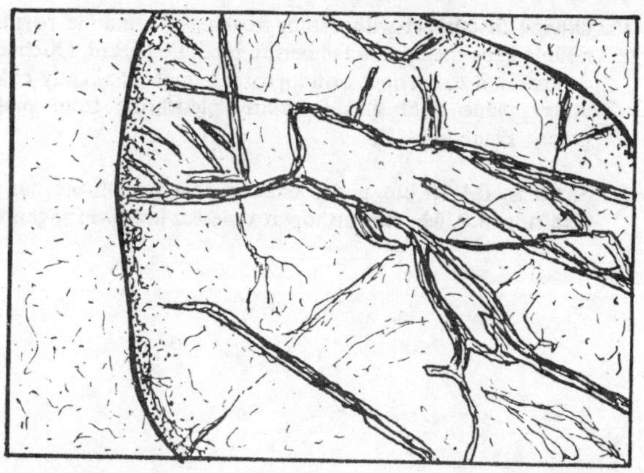

IGNEOUS ROCK-FORMING MINERALS

13. Some olivine, especially in alkaline igneous rocks, contains oriented plates of spinel, which have a peculiar texture resembling a comb or root system that is restricted to a planar direction. This sample is from the Cretaceous alkaline peridotite of Mount Bruno, Quebec. Crossed polars, X80.

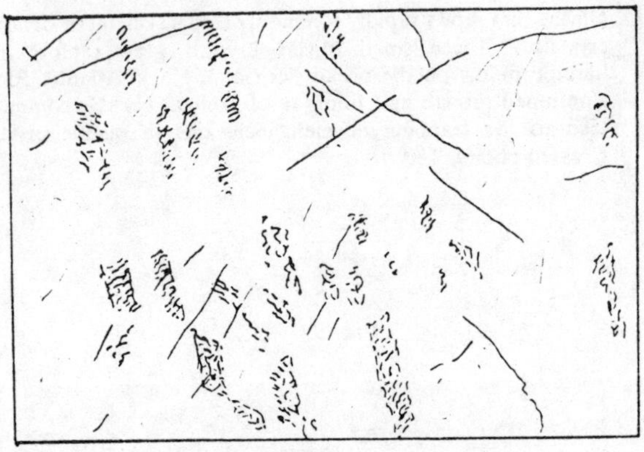

14. Deformed olivine commonly has lamellae parallel to (100) which are visible under crossed polars because of slight differences in crystallographic orientation across lamellae boundaries. Mantle-derived olivine brought to the surface in nodules by alkali basalt commonly exhibit these lamellae, as seen in this example from an 1801 flow from the Hualalai volcano, Hawaii. Crossed polars, X8.

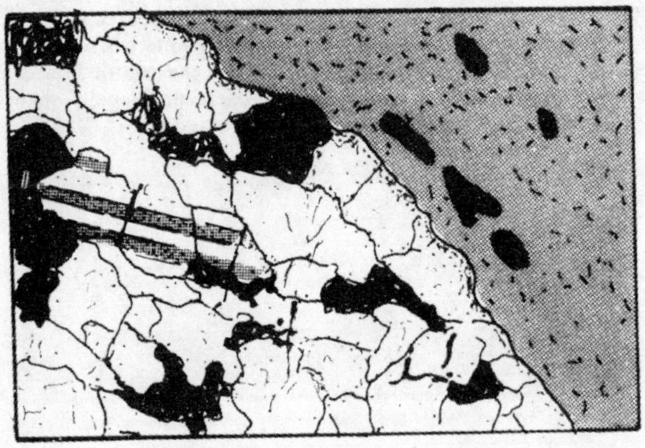

15. Monticellite has the same crystal form as olivine, as shown by this phenocryst in an alnoite from Como, Quebec. Unlike olivine, monticellite has low birefringence. Monticellite occurs only in silica-poor rocks and is commonly associated with melilite (high relief, gray interference colors, 90° cleavage) and nepheline (low relief, in groundmass) but never feldspar. This rock also contains biotite. Crossed polars, X32.

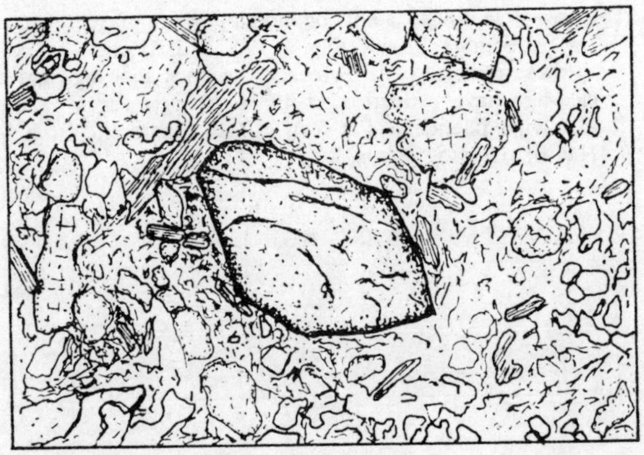

16. Olivine in contact with orthopyroxene in a mantle peridotite nodule brought up in a kimberlite pipe, Ile Bizard, Quebec. In comparison to olivine, orthopyroxene appears slightly colored under plane light and is faintly pleochroic from pink to green. Plane light, X8.

17. Same as IM 16 but under crossed polars. Olivine has high birefringence, whereas orthopyroxene has low birefringence.

IGNEOUS ROCK-FORMING MINERALS

PYROXENES

18. Compositions of nonsodic pyroxenes can be plotted in the lower half of a triangular diagram which has wollastonite at the top and enstatite and ferrosilite at the base; this is known as the pyroxene quadrilateral. Pyroxenes with compositions between diopside and augite, which includes the common augites, are monoclinic and are referred to as clinopyroxenes (Cpx). The calcium-poor pyroxenes are orthorhombic (orthopyroxene = Opx), except at high temperatures where they too are monoclinic. Pigeonite is a relatively calcium-poor clinopyroxene.

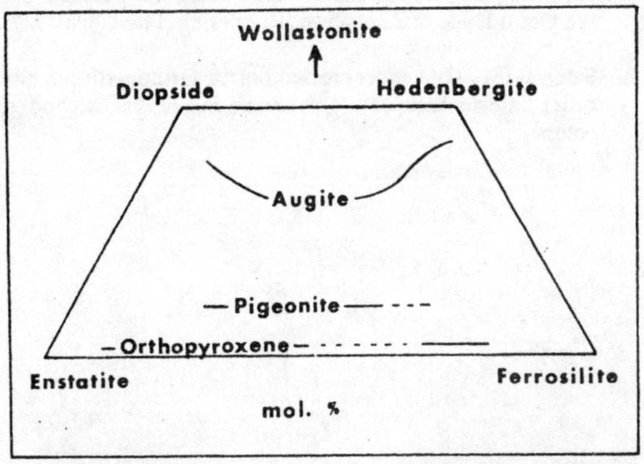

19. Orthopyroxene surrounding olivine grains in a chromite-bearing harzburgite from the Stillwater Complex, Montana. Crossed polars, X8.

20. Orthopyroxene can be distinguished from augite by its low birefringence, as can be seen in this gabbro from the Great Dyke of Zimbabwe (Rhodesia). Augite grains that are oriented so as to show two cleavage directions (center of field) do, however, have low interference colors similar to those of orthopyroxene (top center). Orthopyroxene has extinction parallel to the crystallographic axes, and thus sections having the c axis on the stage of the microscope will extinguish parallel to the cleavage. If the c axis is inclined to the stage the extinction is at an angle to the cleavage, reaching a maximum of 45°. Augite has inclined extinction (~45°), but when viewed down the (010) plane it extinguishes parallel to the cleavage traces. Carefully examine the figures below. Crossed polars, X20.

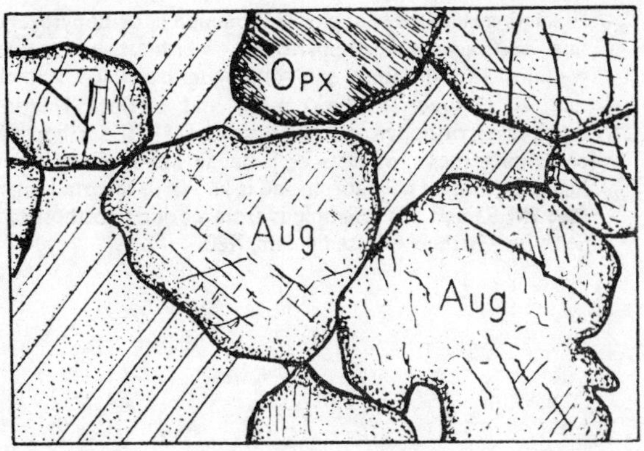

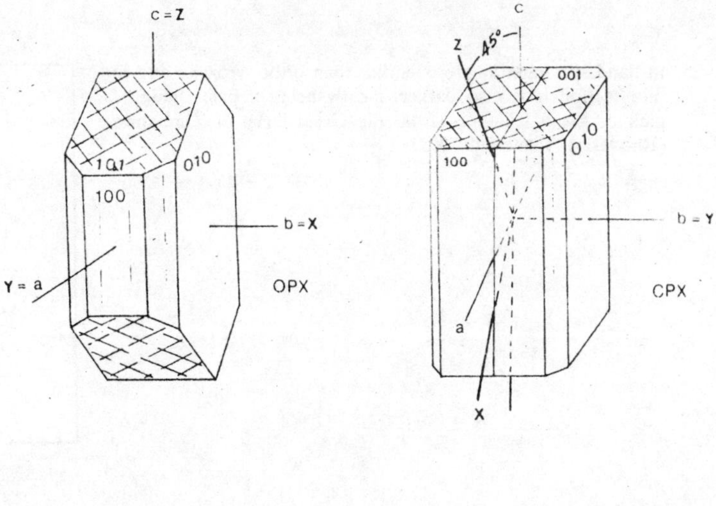

21. Under plane light orthopyroxene is normally clearer than coexisting augite, as seen in this feldspathic pyroxenite from the Great Dyke of Zimbabwe (Rhodesia). Plane light, X20.

22. Same as IM 21 but under crossed polars. Orthopyroxene has first order interference colors, whereas augite has second order colors.

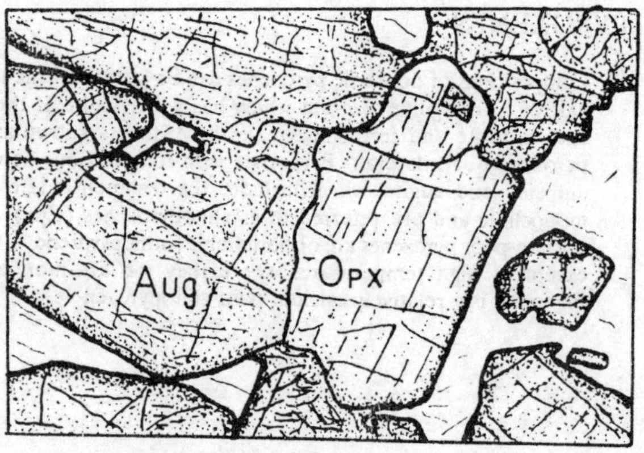

23. Orthopyroxene in the bronzite range (En 88-70%) typically exhibits a lamellar structure parallel to (100). Extinction in these lamellae is a few degrees from being parallel. Some lamellae are produced by exsolution of augite approximately parallel to (100). Thin plates of ilmenite (brown where very thin) may exsolve in this same direction, which gives this pyroxene its bronze appearance in hand specimen. Great Dyke of Zimbabwe (Rhodesia). Crossed polars, X20.

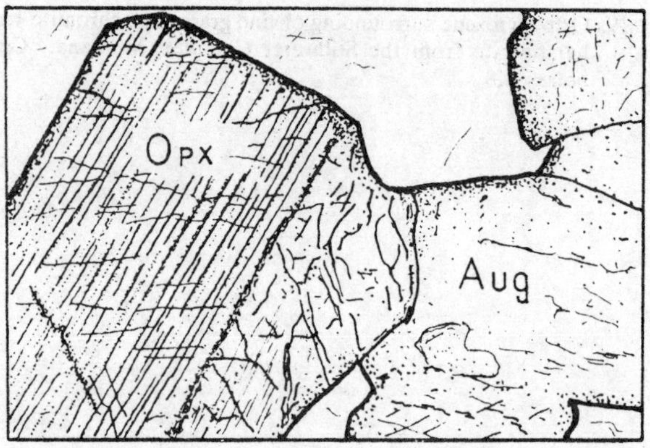

24. Exsolution lamellae of augite are common in orthopyroxenes. In this section the orthopyroxene, which is in extinction, contains two different sets of augite lamellae, a thick set that is oblique to the c axis, and a finer set in the intervening orthopyroxene that parallels (100). Orthopyroxene containing these two sets of lamellae has inverted from the high-temperature polymorph, and is known as inverted pigeonite (see IM 32 and 33). Sample is from noritic anorthosite, Lake St. John, Quebec. Crossed polars, X20.

25. In hand specimen augite is darker than orthopyroxene and in thin section it is also darker, mostly being a pale shade of pink or green. Gabbro from the Great Dyke of Zimbabwe (Rhodesia). Plane light, X20.

26. Augite that contains significant amounts of titanium is distinctly darker than ordinary augite, and is colored in tints of brownish, purplish, or violet. Titanium-rich augite occurs only in silica-poor rocks, as in this essexite (nepheline gabbro) from Mount Yamaska, Quebec. Plane light, X20.

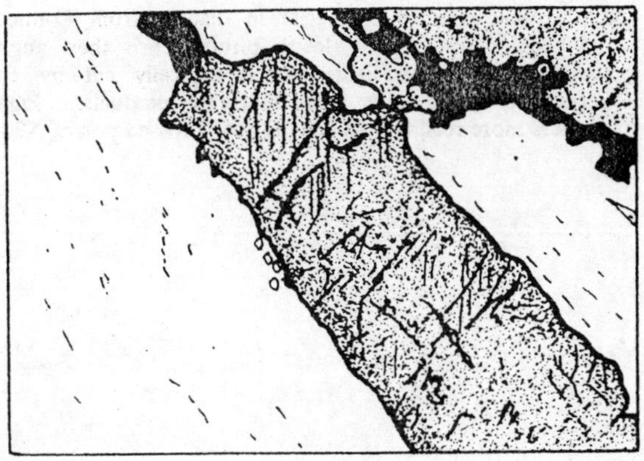

27. In alkaline rocks, clinopyroxenes are commonly zoned. In this example from Scawt Hill, Northern Ireland, augite is zoned outward through a purplish brown titanaugite to a rim of green aegerine. The groundmass contains nepheline. Plane light, X20.

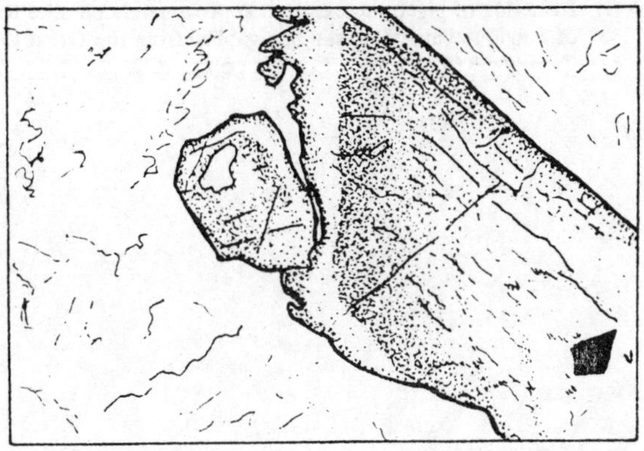

28. Aegerine-augite zoned to rims of aegerine in nepheline syenite. Also present are zoned crystals of sodic hornblende. The amphibole is distinctly less transparent than the pyroxene. Hexagonal prisms of apatite are also present. Mount Johnson, Quebec. Plane light, X80.

29. Orthopyroxene crystals poikilitically enclosed in plagioclase and augite. Both the orthopyroxene and augite contain exsolution lamellae of the other parallel to (100). Stillwater Complex, Montana. Crossed polars, X8.

30. Pigeonite rimmed by augite in diabase from Connecticut. Pigeonite has slightly lower birefringence than augite, is clearer in plane light, and is typically cut by curving fractures filled with dark alteration products. Pigeonite alters more readily than does augite. Crossed polars, X20.

31. Lamellae of pigeonite exsolved approximately parallel to (001) of a twinned augite crystal in a gabbro from the Great Dyke of Zimbabwe (Rhodesia). Crossed polars, X32.

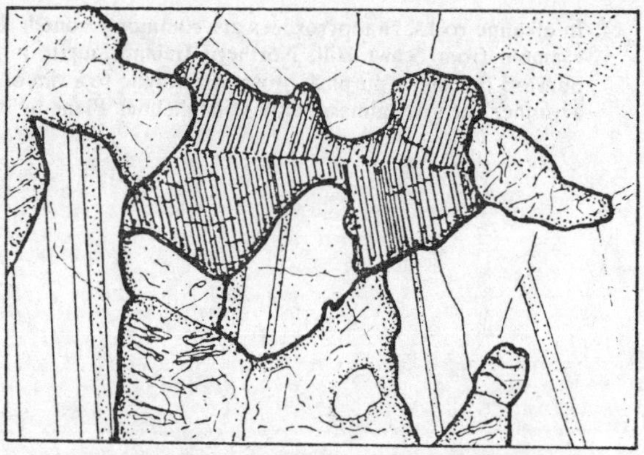

32. Large single crystal of orthopyroxene which inverted from pigeonite. Exsolution of augite from the original twinned pigeonite crystal was approximately parallel to (001) of pigeonite. The twin must have been present in the original monoclinic pigeonite, because twinning is not possible in this direction in orthorhombic pyroxene. Note that the cleavage in the orthopyroxene passes straight through the boundary of the twin. The original twin is therefore only made evident by the orientation of the exsolution lamellae. In the upper left, an augite grain contains (001) lamellae of pigeonite. Great Dyke of Zimbabwe (Rhodesia). Crossed polars, X80.

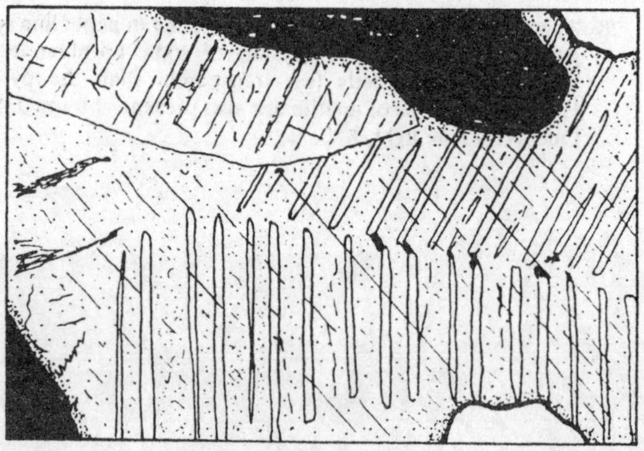

33. Complex pattern of exsolution lamellae indicating that originally a crystal of pigeonite was rimmed by augite; exsolution took place in both the pigeonite and augite approximately parallel to (001)--the horizontal sets; then the pigeonite inverted to orthopyroxene; finally, exsolution in both the orthopyroxene and augite took place parallel to (100)--fine vertical sets. Orthopyroxene seen here in the extinction position. Lake St. John anorthosite massif, Quebec. Crossed polars, X20.

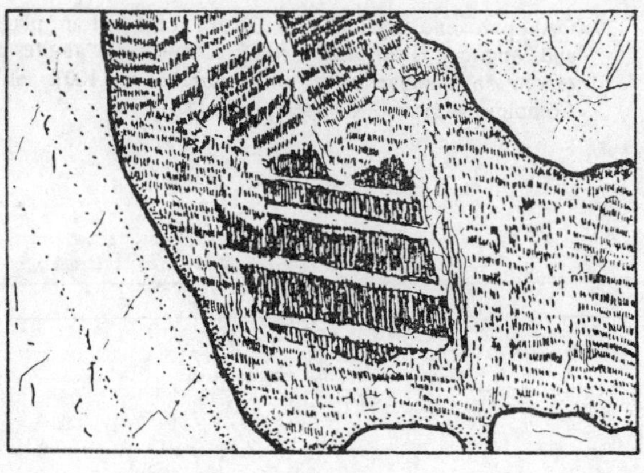

IGNEOUS ROCK-FORMING MINERALS

AMPHIBOLES

34. In hand specimen hornblende is normally black and cleavage surfaces give bright reflections. Pyroxene, by contrast is brownish or greenish and poorer cleavage gives less reflective surfaces. Here phenocrysts of hornblende, up to a centimeter across, occur in a camptonite dike. Hornblende crystals in mafic rocks that are erupted as lava or intruded to shallow depth commonly are rounded due to resorption. Montreal, Quebec.

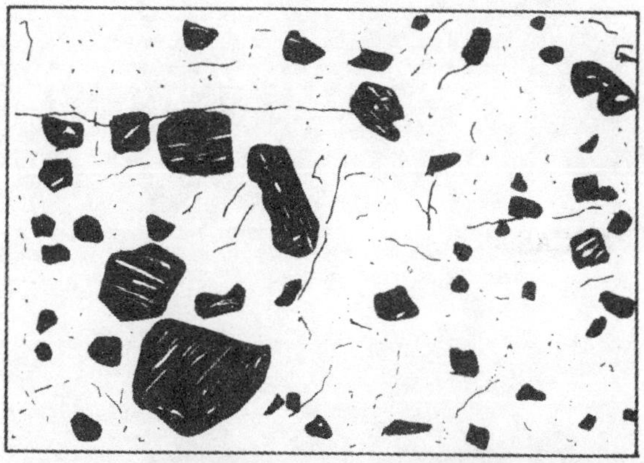

35. Hornblende crystals commonly have simple twins. Hornblende is strongly pleochroic, as seen here in the two halves of the twinned crystal. These euhedral hornblende phenocrysts occur with ragged olivine phenocrysts in a lamprophyric dike, Mount Bruno, Quebec. Plane light, X20.

36. Same as IM 35 but under crossed polars. Hornblende shows second order interference colors, but these may be masked by the strong color of the mineral.

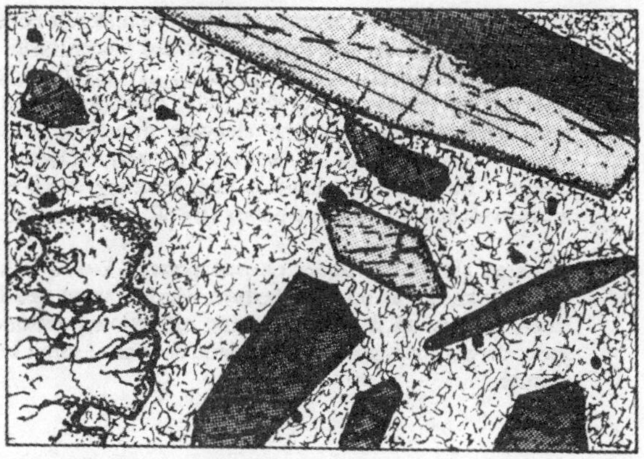

37. Basal sections of hornblende exhibit characteristic 56-124° amphibole cleavage. Sections which do not exhibit pleochroism under plane polarized light give centered optic axis figures under crossed polars. Nepheline syenite, Mount Johnson, Quebec. Plane light, X32.

38. The blue sodic amphibole riebeckite commonly forms acicular crystals, which in this quartz syenite from Grenville, Quebec, form radiating clusters. Plane light, X8.

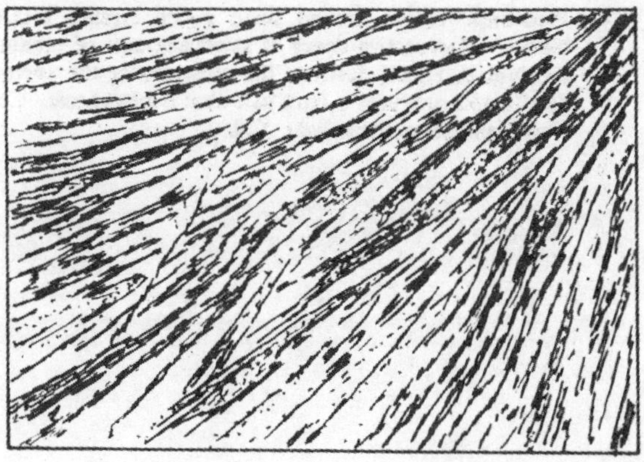

MICAS

39. Biotite flanked by hornblende. These two minerals commonly occur together and have similar color and pleochroism. They can be distinguished by the presence of only one cleavage direction in biotite, the perfect (001) cleavage. Even when hornblende shows only one cleavage direction, as in this section, the cleavage is not as perfectly planar as in biotite. Biotite also exhibits a faint mottling which is made more evident under crossed polars (IM 40). Plane light, X80.

40. Same as IM 39 but under crossed polars. Biotite exhibits a mottling known as birdseye texture, which is most clearly seen when the crystal is rotated to the extinction position--small spots do not extinguish. Hornblende does not exhibit birdseye texture.

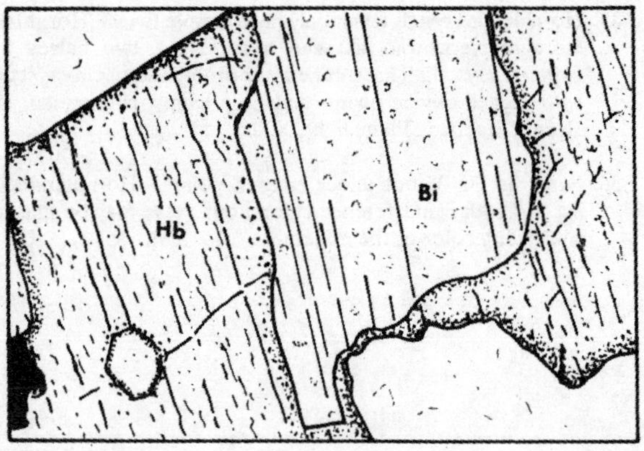

41. Muscovite and bitite in Conway granite, New Hampshire. Muscovite and biotite have similar birefringence and both exhibit birdseye texture. Crossed polars, X20.

42. Same as IM 41 but under plane light. Muscovite is colorless and biotite is brown and pleochroic. The magnesian biotite, phlogopite, which is colorless, can be distinguished from muscovite by its small 2V.

43. Pleochroic haloes around small zircon crystals included in biotite. Radioactive decay of small amounts of uranium and thorium in zircon destroys the structure of the surrounding biotite and produces a dark halo that is pleochroic. Concord, New Hampshire. Plane light, X20.

IGNEOUS ROCK-FORMING MINERALS

FELDSPARS

44. Plagioclase feldspar in most igneous rocks is characterized by multiple twinning. The most common twin law in plagioclase is the albite twin, with (010) twin plane (horizontal). Pericline twins, which have the rhombic section as twin plane, are oriented roughly perpendicular to the albite twins and commonly form short multiple twins between the albite lamellae giving the appearance of rungs on a ladder. Crossed polars, X8.

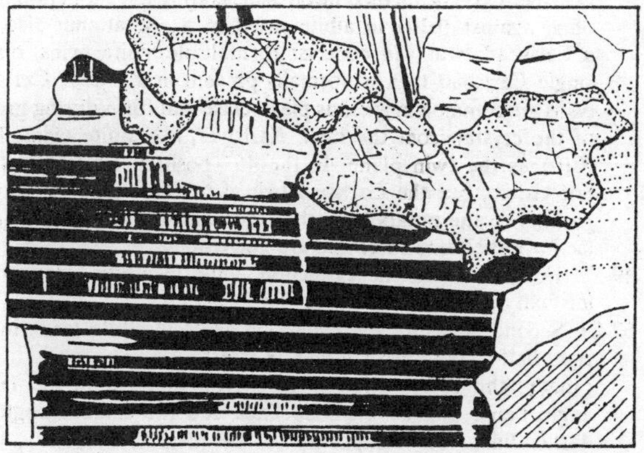

45. The composition of plagioclase can be determined by measuring the orientation of the optical indicatrix with respect to crystallographic directions. This can be done by measuring the extinction of the fast vibration direction (X') against the (010) plane as defined by the albite twins. Measurements must be made on crystals with (010) vertical. Such sections show sharp twin lamellae that are equally illuminated when parallel to the polars; also their image does not shift from side to side when the microscope is focused up and down under high power. Extinction angles measured on the two sets of albite lamellae should then be the same. Measurements are repeated on as many crystals as possible until a maximum value is obtained. With this value the composition of the plagioclase is read from the graph (p. 46). Crossed polars, X8.

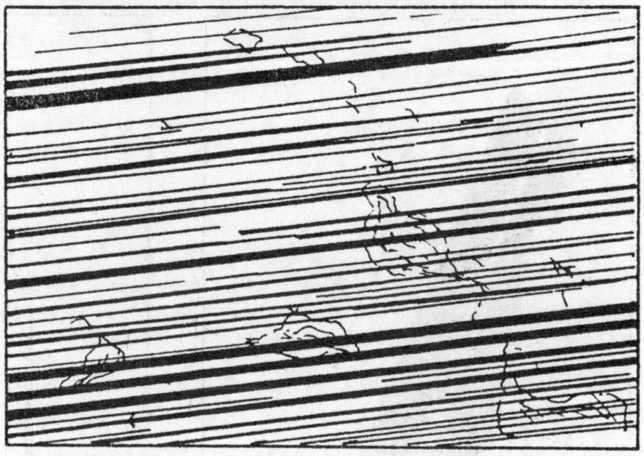

46. Plagioclase compositions can be determined from a single measurement of the extinction angle of X' against (010) if the section is perpendicular to both (010) and (001). Such sections, in addition to having sharp albite twins, show (001) cleavage cutting roughly at right angles across the albite twins, as seen in this section. The cleavage is checked for verticality by focusing the microscope up and down under high power. Crossed polars, X8.

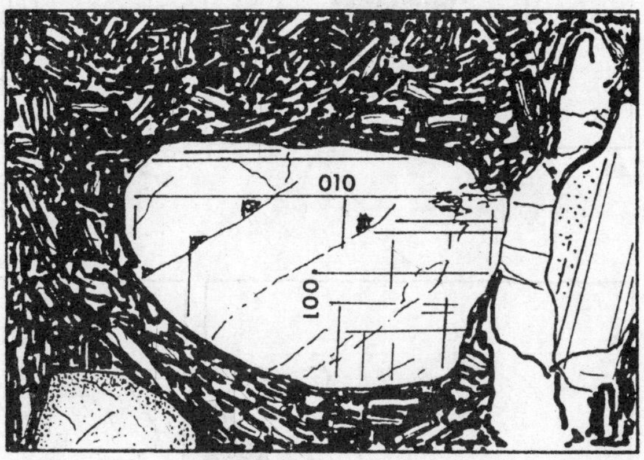

47. Same section as IM 46 but rotated so that one set of twin lamellae is in extinction. The angle between the fast vibration direction (X') and (010) is 41°, which corresponds to a composition of An_{80}. Plagioclase phenocryst in Mid-Atlantic Ridge basalt.

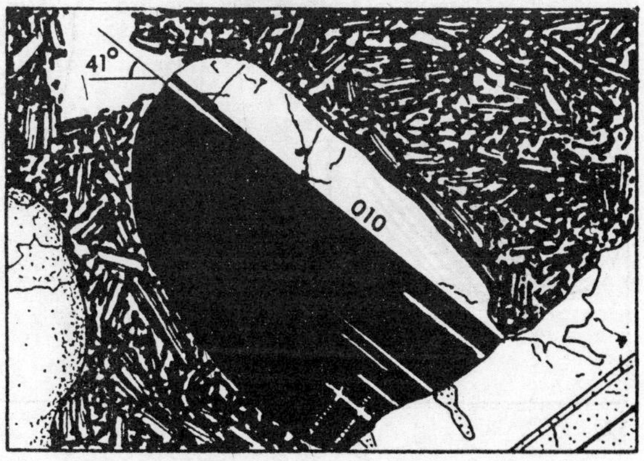

48. Determination of plagioclase composition by Carlsbad--albite twin method. The method involves measuring the extinction angle of x' against (010) in albite twins in a crystal that also has a Carlsbad twin. In contrast to multiple albite twins, only a single Carlsbad twin is normally present in a crystal. Carlsbad twins are common in igneous plagioclase and form during growth of the crystal from the melt. Carlsbad, like albite twins, have (010) as the twin plane. In the 45° position albite twins are not visible, but the Carlsbad twin is. When {010} is rotated parallel to the polars both the albite and Carlsbad twin lamellae have the same color and are not visible.

49-51. First, measure the extinction angle in albite twins in the left half of the Carlsbad twin (= 24).

52-53. Second, measure the extinction angle in albite twins in the right half of the Carlsbad twin (= 40).
Finally, the composition is read from the graph, the larger angle being plotted on solid lines and the smaller angle on dashed lines. The composition of this plagioclase is An_{72}.

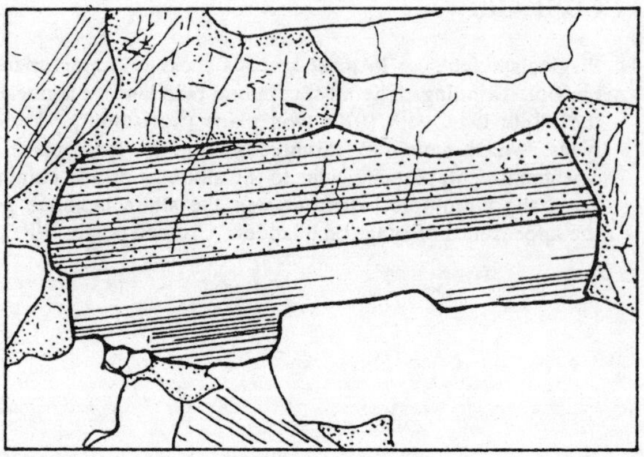

48.

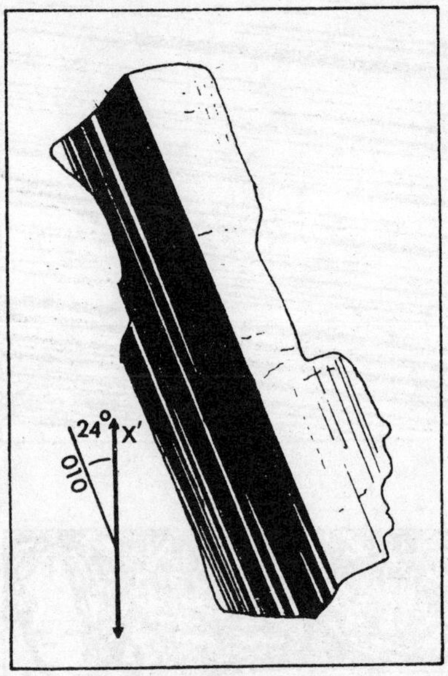

49.

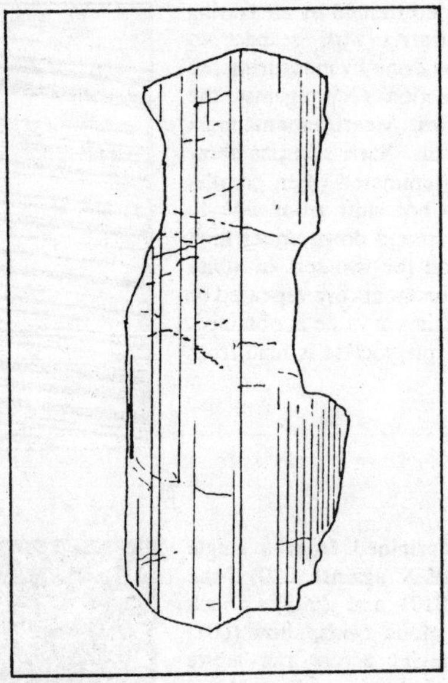

50.

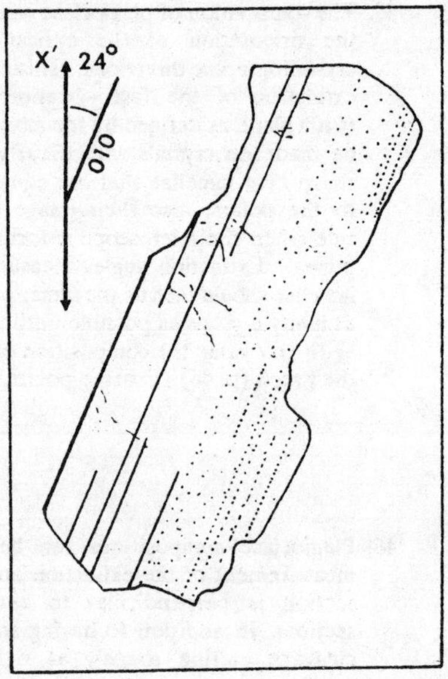

51.

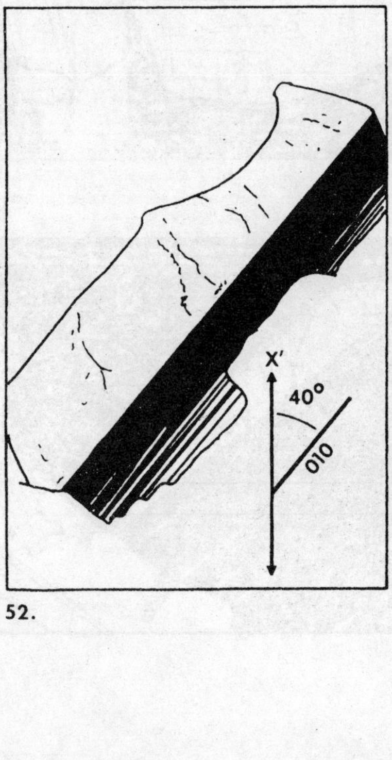

52.

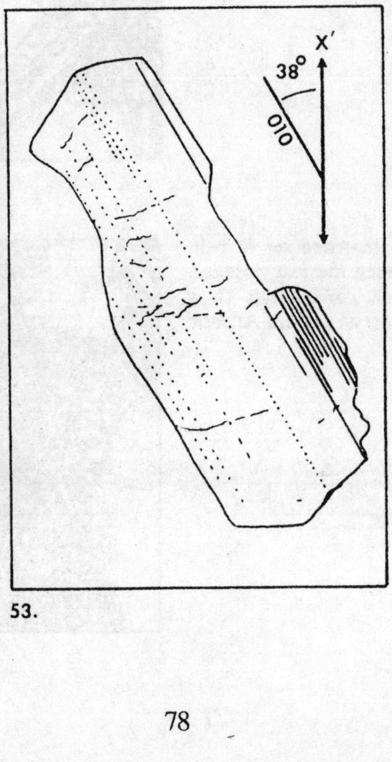

53.

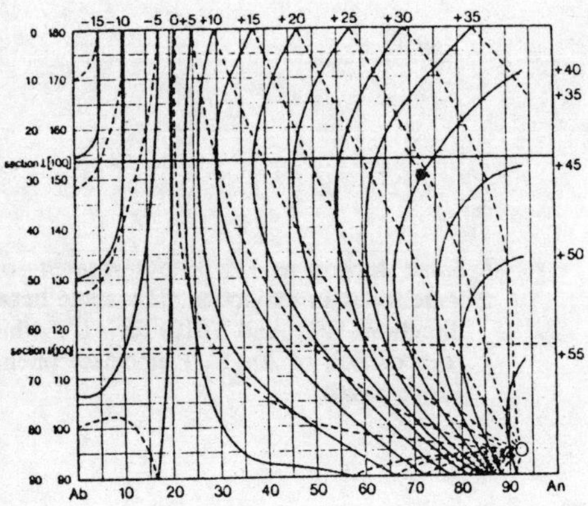

54. Compositional zoning in plagioclase is common. If the crystal is more sodic toward the rim, it is said to be normally zoned; if it is more calcic, it is reversely zoned. If the composition fluctuates back and forth, the plagioclase has oscillatory zoning, as seen in this phenocryst in an andesite from Mount St. Helens. Crossed polars, X20.

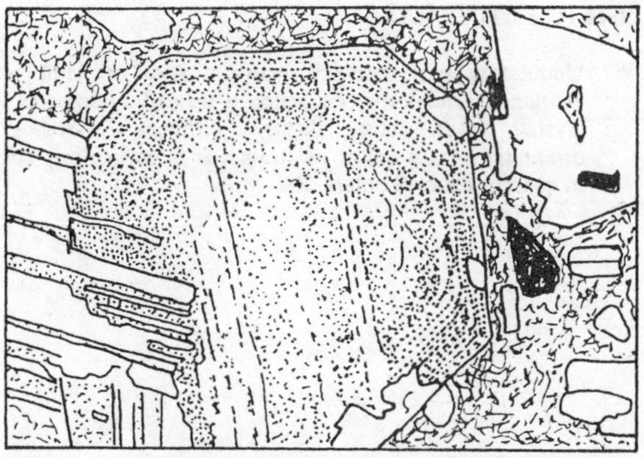

55. When plagioclase grows rapidly, as in the groundmass of this olivine basalt from the Mid-Atlantic Ridge, it forms dendritic microlites that are elongated parallel to crystallographic a. The composition of these microlites can be determined by measuring the maximum extinction angle of X' against the length of the crystals (p. 46). Plane light, X80.

56. Orthoclase perthite formed by exsolution of albite-rich feldspar from potassium-rich feldspar. Crossed polars, X20.

57. Same as IM 56 but under plane light. The potassium feldspar host is seen to have lower relief than the albite lamellae or the quartz.

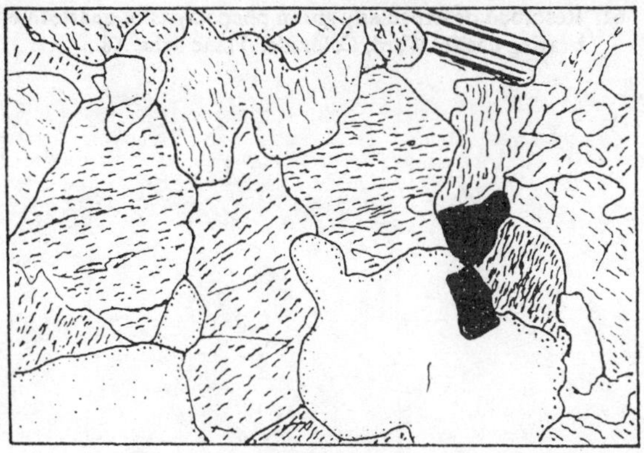

58. Grid twinning, which is characteristic of most microcline, results from extremely fine albite and pericline twins that form when the feldspar inverts, on cooling, from the higher-temperature monoclinic orthoclase to the lower-temperature triclinic microcline. Crossed polars, X20.

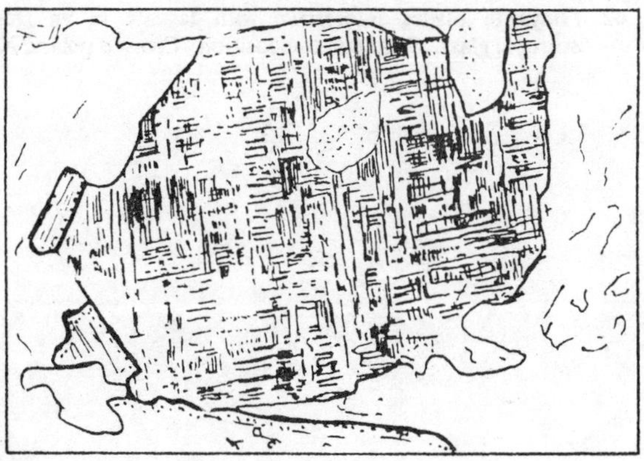

IGNEOUS ROCK-FORMING MINERALS

SILICA MINERALS

59. Undulatory extinction in quartz. Ross of Mull granite, Scotland. Undulatory extinction is formed by domains within crystals that have slightly different crystallographic orientations as a result of strain. It is particularly common in quartz. Crossed polars, X8.

60. Quartz phenocrysts with a dipyramid form indicating that they crystallized as the high-temperature beta form in this rhyolite from Salem, Massachusetts. Crossed polars, X8.

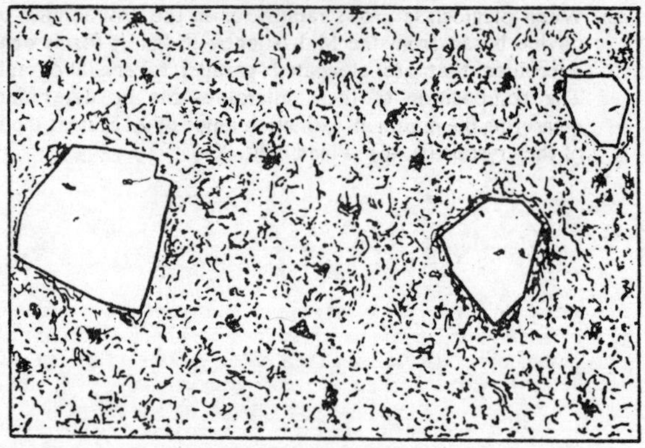

61. Resorbed or irregularly grown phenocrysts of quartz in biotite-bearing dacite, Ward, Colorado. Plane light, X8.

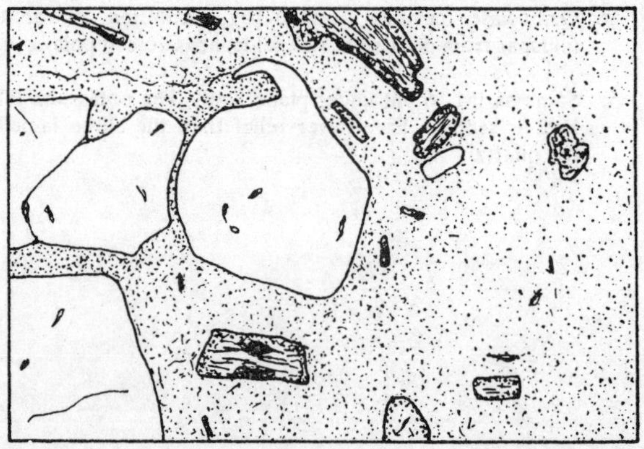

62. Tridymite blades intergrown with fayalite in an iron-rich isotropic glass. Synthetic composition. Crossed polars, X25.

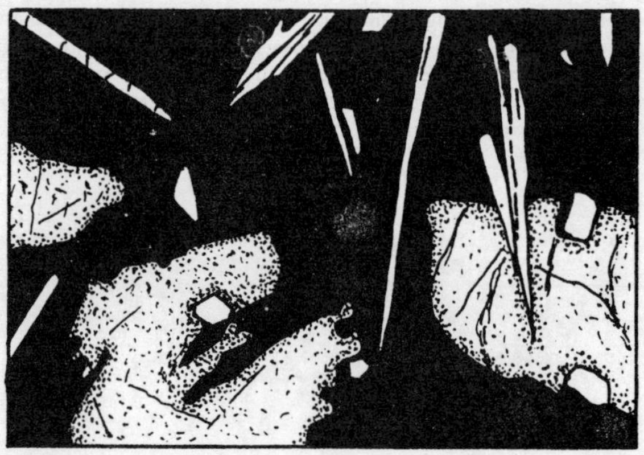

IGNEOUS ROCK-FORMING MINERALS

FELDSPATHOIDS

63. Late-crystallizing nepheline surrounding plagioclase and hornblende in essexite, Mount Johnson, Quebec. Nepheline has extremely low birefringence–first order gray. Crossed polars, X8.

64. Same as IM 63 but under plane light. Nepheline alters readily, as seen from its clouded appearance. By contrast, quartz, which also commonly crystallizes late and thus exhibits similar textural relations to those of nepheline, shows no alteration.

65. Because of its low birefringence, nepheline goes to extinction easily and when in this position shows an abundance of small bright flecks of muscovite formed from alteration of the nepheline. Essexite, Mount Johnson, Quebec. Crossed polars, X20.

66. Phenocrysts of nepheline in phonolite, Mount St. Hilaire, Quebec. Dark phenocrysts are oxidized fayalite. Plane light, X8.

67. Same as IM 66 but under crossed polars. Those sections of nepheline phenocrysts with hexagonal outline have low interference colors, whereas those with squarish outline show higher white interference colors.

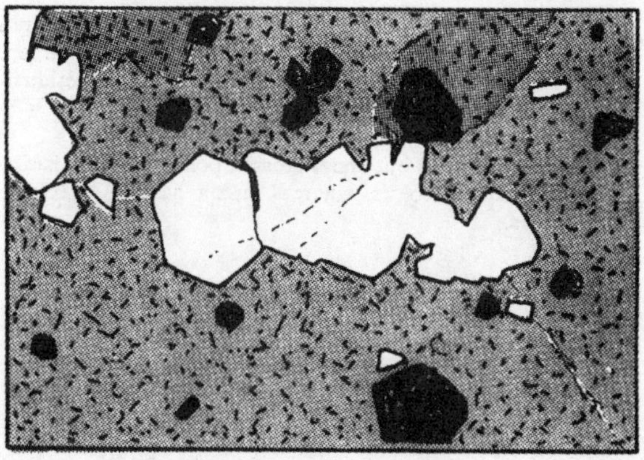

68. Euhedral phenocrysts of leucite and aegerine in a leucitite, Vesuvius. Plane light, X20.

69. Same as IM 68 but under crossed polars. Leucite, on cooling, inverts from cubic to tetragonal, and multiple twinning develops in different sectors.

IGNEOUS ROCK-FORMING MINERALS

70. Bladed zeolite, thomsonite, with isotropic analcite filling vesicles in basalt, Island of Mull, Scotland. Crossed polars, X8.

71. Melilite, nepheline, titanaugite, and magnetite in nephelinite from the Honolulu series, Oahu, Hawaii. Melilite and nepheline both have low gray interference colors. They can easily be distinguished by relief--melilite is much higher. Crossed polars, X80.

72. Same as IM 71 but under plane light where the low relief of nepheline is evident.

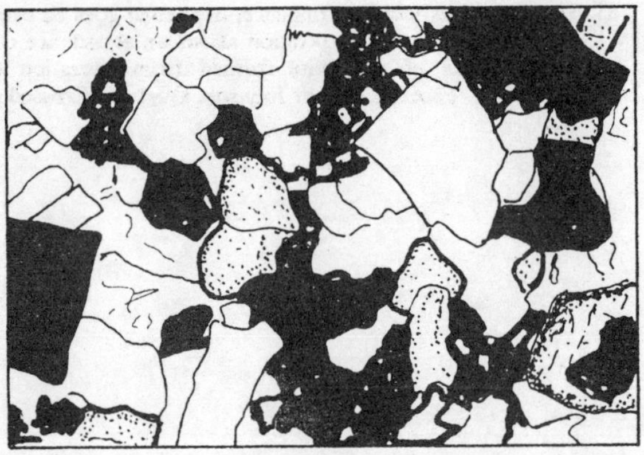

73. Slender prisms of apatite--one with an axial cavity--and a square section through a zircon crystal beside a larger biotite crystal. Apatite crystallizing in silica-rich residual melts commonly forms long needles. Plane light, X80.

74. Same as IM 73 but under crossed polars. Apatite has low gray interference colors, whereas zircon has up to fourth order colors.

75. In mafic rocks, apatite typically forms stubby prisms, as seen here enclosed in hornblende, augite, and magnetite in an essexite, Mount Johnson, Quebec. Although apatite does not have cleavage, prominent fractures across the prisms are common.

76. Same as IM 75 but under crossed polars.

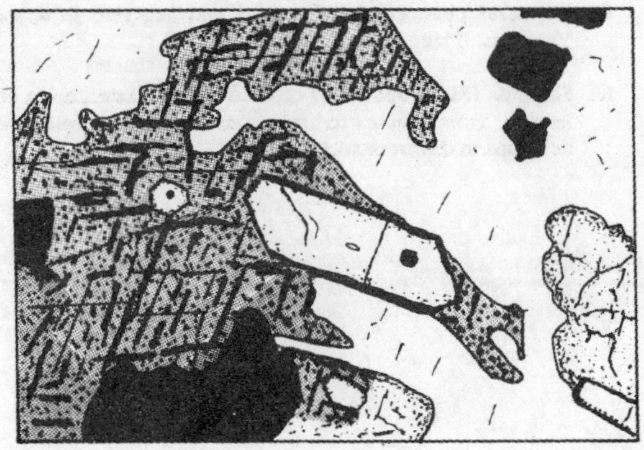

IGNEOUS ROCK-FORMING MINERALS

77. Typical wedge-shaped crystal of sphene showing the prominent cleavage. In nepheline syenite, Loch Borrolan, Scotland. Plane light, X80.

78. Same as IM 77 but under crossed polars. Sphene appears essentially the same under crossed polars as under plane light, because its high order pink interference color is masked by the mineral's normal color.

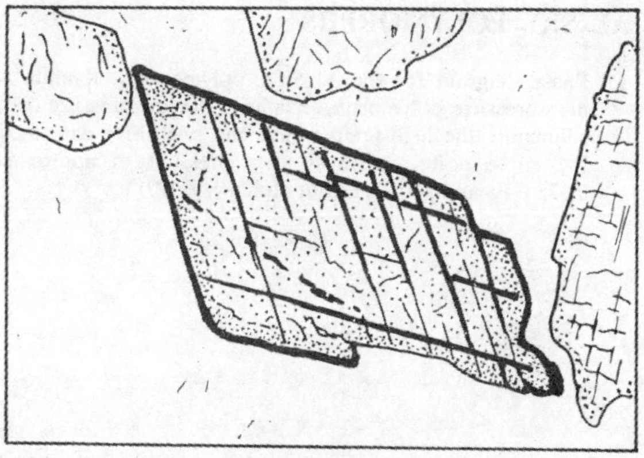

79. Zoned tourmaline crystals with greenish cores and brownish rims. Plane light, X20.

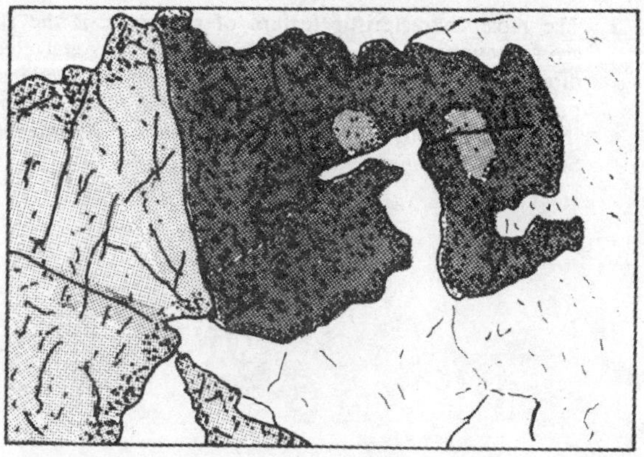

80. Alkali gabbro containing titanaugite, plagioclase, and magnetite, which encloses dark green hercynite. Mount Yamaska, Quebec. Plane light, X80.

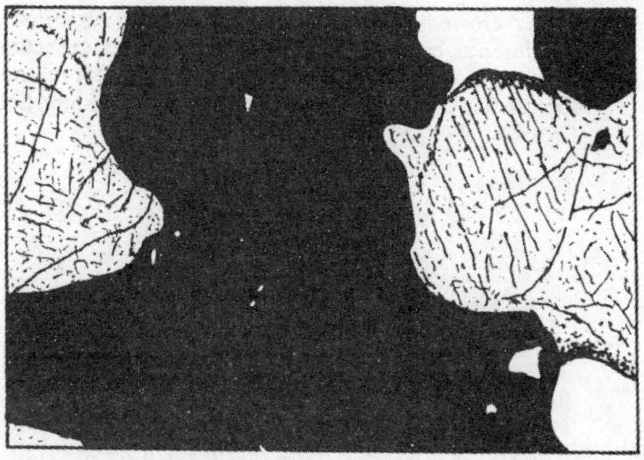

METAMORPHIC ROCK-FORMING MINERALS

AL_2SiO_5 POLYMORPHS

1. Phase diagram for the Al_2SiO_5 polymorphs. Kyanite is the high-pressure polymorph, andalusite the low-pressure one, and sillimanite the high-temperature one. All three can coexist at the triple point, which is at a pressure of approximately 0.375 GPa and 500°C. (After Holdaway, 1971)

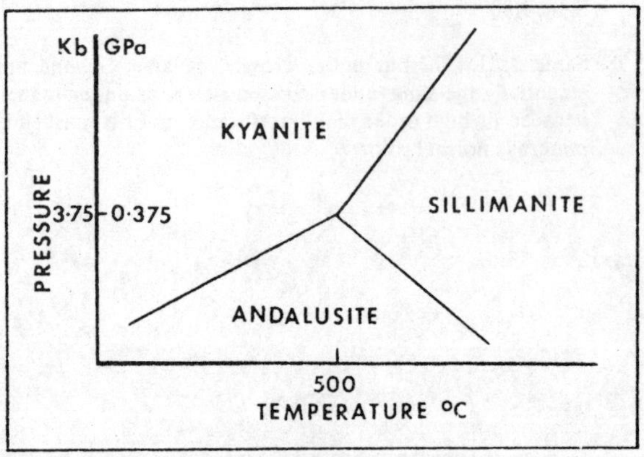

2. The most characteristic feature of sillimanite is the square cross section of the prism with a single diagonal cleavage direction. Cleavage is not evident in prismatic sections, but prominent cross fractures are common. Country rock surrounding the mafic igneous complex of Haddo House, Aberdeenshire, Scotland. Plane light, X30.

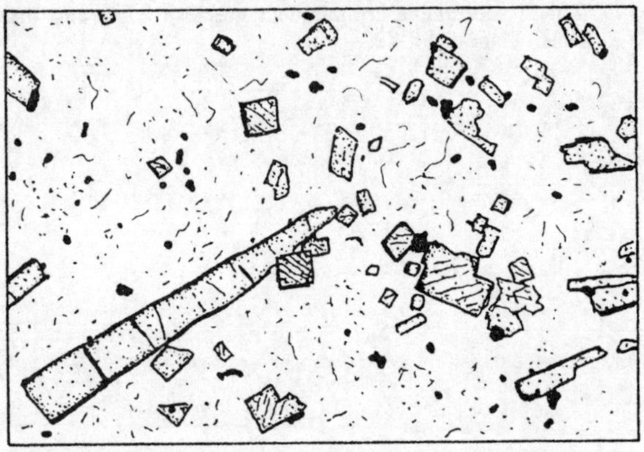

3. Under crossed polars, sillimanite has up to high first order interference colors, and prisms have parallel extinction and are length slow. Seen here with cordierite from the hornfels surrounding the Haddo House Complex, Aberdeenshire, Scotland. Crossed polars, X10.

4. Sillimanite can occur as a fibrous mat known as fibrolite, seen here beside a garnet crystal. The fibers have the same optical properties as the coarser prisms, but they are commonly so fine that individual fibers cannot easily be resolved in the optical microscope. Crossed polars, X54.

METAMORPHIC ROCK-FORMING MINERALS

5. Kyanite commonly forms coarse bladed crystals. Although it is essentially colorless in thin section, it appears quite dark because of its high refractive indices. Country rock surrounding Ross of Mull granite, Scotland. Plane light, X9.

6. Same as MM 5 but under crossed polars. Kyanite has upper first order interference colors.

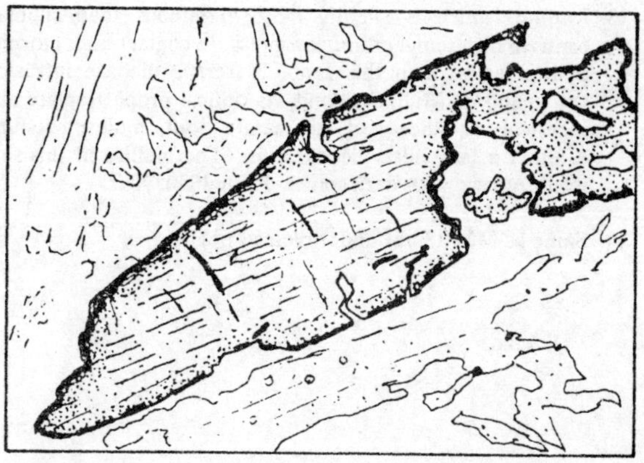

7. Kyanite exhibits two cleavages that parallel the length of the blades and intersect at a high angle. A prominent cross fracture is also evident in prismatic sections. Ross of Mull, Scotland. Plane light, X20.

8. Same as MM 7 but under crossed polars. Blades of kyanite are length slow, and the extinction angle varies from 0 to 32° depending on the orientation of the crystal.

9. Andalusite typically forms large porphyroblasts with poorly developed crystal faces. It has the lowest refractive indices of the three Al_2SiO_5 polymorphs. In this section a large andalusite crystal is beside several smaller, higher relief crystals of kyanite. This andalusite formed in the contact metamorphic aureole around the Ross of Mull granite. Kyanite was the stable polymorph in the regionally metamorphosed rock prior to the intrusion of the granite. Plane light, X10.

10. Same as MM 9 but under crossed polars. First order yellow is the maximum interference color of andalusite.

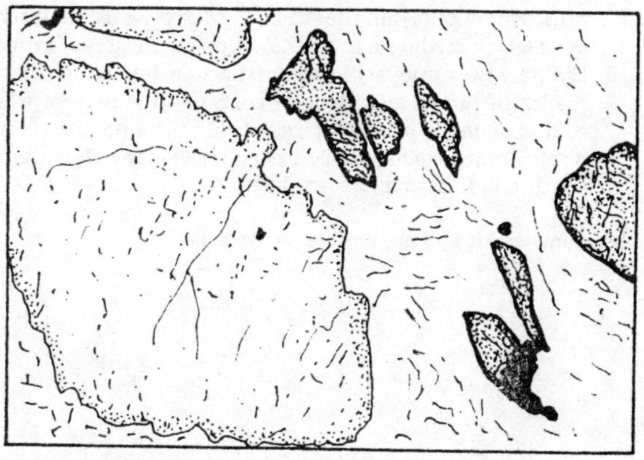

11. Porphyroblasts of andalusite commonly grow around and enclose many of the other minerals in a rock. These included minerals, especially graphite, may be concentrated at the core or along diagonal planes through the prism. This latter arrangement produces a cruciform pattern of inclusions in sections parallel to the basal plane of the prism. This variety of andalusite is known as chiastolite. Plane light, X4.

12. Same as MM 11 but under crossed polars.

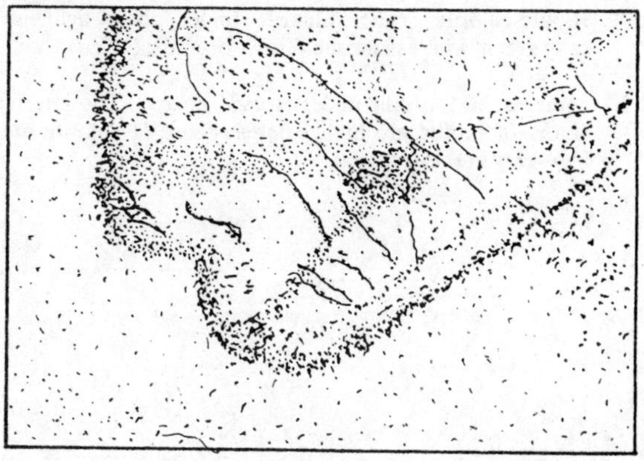

METAMORPHIC ROCK-FORMING MINERALS

13. Mullite, which is slightly more aluminous than sillimanite, forms in high-temperature, low-pressure contact metamorphosed pelites, as seen in this xenolith from a diabase intrusion in the Island of Mull, Scotland. Its optical properties are almost identical to those of sillimanite, but mullite commonly exhibits a faint lilac pleochroism. The mullite in this sample is intergrown with dark spinel. Plane light, X80.

14. Same as MM 13 but under crossed polars.

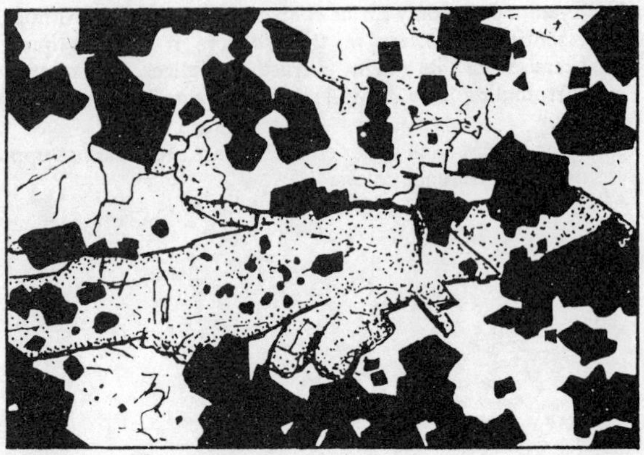

15. Corundum, seen here intergrown with plagioclase, dark euhedral spinel, and dark brown interstitial glass, is also a common constituent of high-temperature contact metamorphosed pelites. Corundum forms euhedral stubby prisms with extremely high relief. It has no cleavage, but a poor basal parting produces transverse fractures in prismatic sections. Xenolith in diabase, Mull, Scotland. Plane light, X20.

16. Same as MM 15 but under crossed polars. The birefringence of corundum is the same as that of quartz, but because corundum is so hard it is normally thicker than the other minerals in the section and will, therefore, have slightly higher interference colors.

17. Muscovite-biotite-chlorite schist. Muscovite is characterized by high birefringence and a stippled birdseye extinction. Biotite also exhibits birdseye extinction but is pleochroic in shades of brown and green in contrast to muscovite which is colorless under plane-polarized light. Chlorite has very low birefringence, and typically has anomalous interference colors --brown in this sample. Plane light, X8.

18. Same as MM 17 but under crossed polars.

19. Biotite-chlorite schist. Chlorite under plane light is typically green, whereas most biotite is brownish.

20. Same as MM 19 but under crossed polars. The interference colors of biotite are high, whereas those of chlorite are very low--gray in this sample.

METAMORPHIC ROCK-FORMING MINERALS

21. Garnet porphyroblast in a muscovite schist. In pelitic schists garnet typically forms euhedral porphyroblasts. Small amounts of brown biotite and pale green chlorite are also present. Garnet in pelitic rocks belongs to the almandine (Fe) -- spessartine (Mn) -- pyrope (Mg) series and is colored some shade of pale pink in thin section. Plane light, X8.

22. Same as MM 21 but under crossed polars. Garnet in pelitic rocks is isotropic, but calcium-bearing garnets in metamorphosed limestones and skarns typically show anomalous low interference colors.

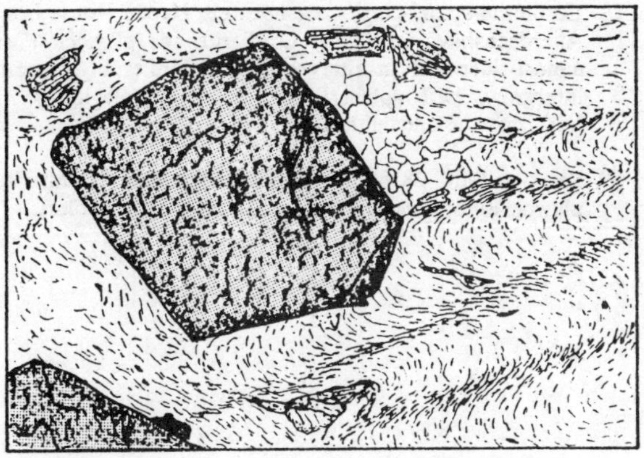

23. Titanium-rich garnet, melanite, is dark brown in plane light. X20.

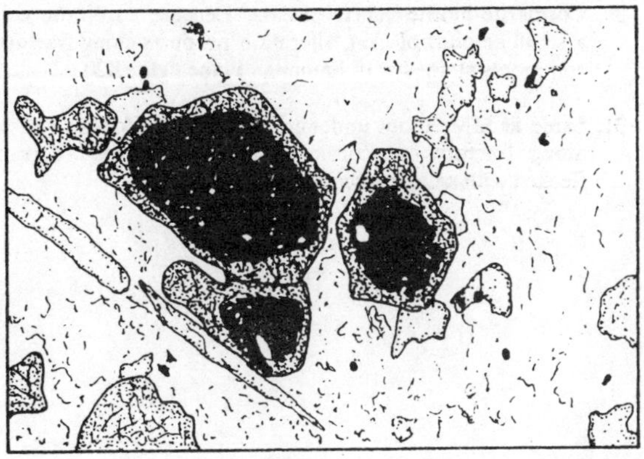

24. Euhedral staurolite porphyroblasts beside a large anhedral garnet porphyroblast in a muscovite-graphite schist. Staurolite is one of the few minerals that is pleochroic in shades of yellow. Plane light, X8.

25. Same as MM 24 but under crossed polars. The maximum interference color of staurolite is first order yellow.

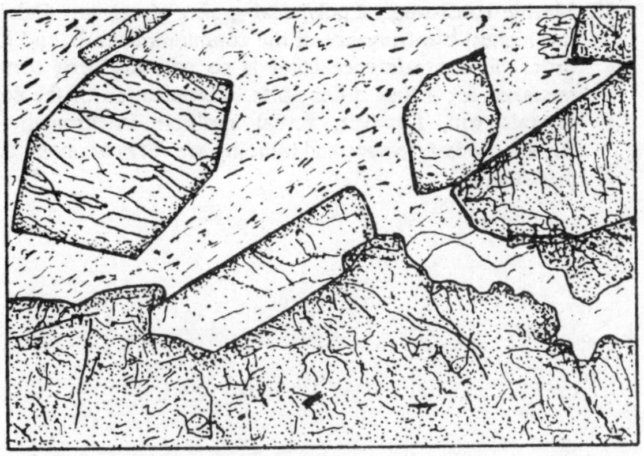

26. Chloritoid invariably grows as porphyroblasts in iron-rich pelitic rocks at low grades of metamorphism. Crystals commonly cut across the foliation defined by muscovite. The pleochroism, from grayish green through blue to pale yellow, is diagnostic of chloritoid. Unlike in micas, fractures do cut across the basal cleavage. Twinning on the basal plane is common. Plane light, X20.

27. Same as MM 26 but under crossed polars. Because of low, anomalous interference colors, the color of chloritoid under crossed polars does not differ much from that under plane light.

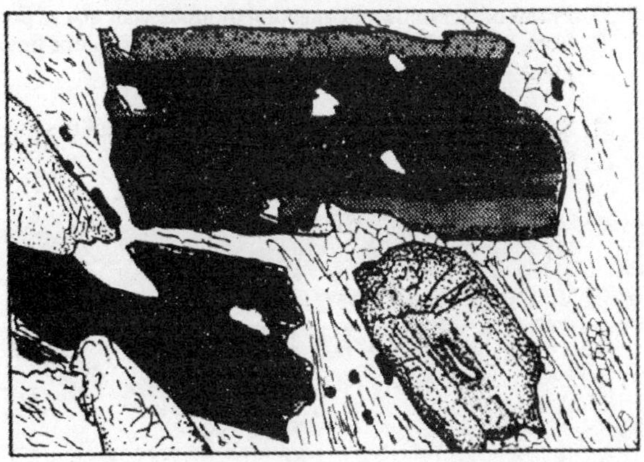

METAMORPHIC ROCK-FORMING MINERALS

28. Cordierite-spinel-plagioclase hornfels, Haddo House Complex, Aberdeenshire, Scotland. Cordierite and plagioclase appear almost identical in thin section. Cordierite is commonly intergrown with spinel (upper right) but plagioclase (lower left) is not. Plane light, X20.

29. Same as MM 28 but under crossed polars. Cordierite and plagioclase both have gray to white interference colors.

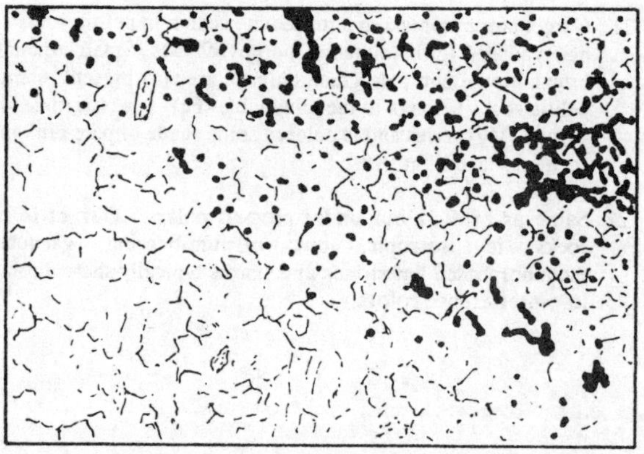

30. Cordierite-biotite-quartz gneiss. Because cordierite contains a small amount of iron, alteration products along fractures are commonly stained with limonite. Plane light, X20.

31. Same as MM 30 but under crossed polars. Zones of alteration along fractures in cordierite appear as broad dark zones flecked with small muscovite crystals (pinite).

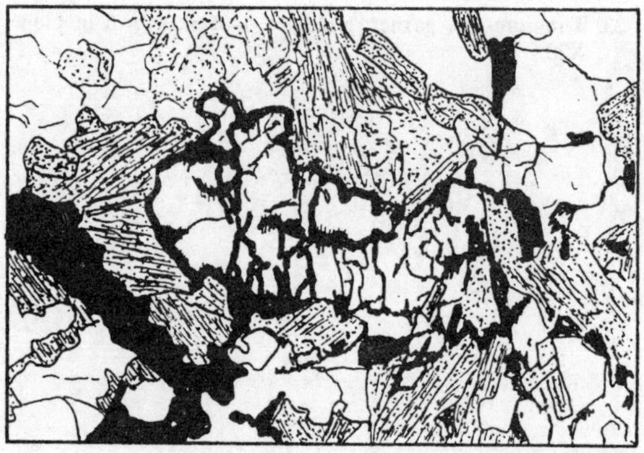

32. Marble consisting of polygonal grains of calcite. Calcite has the highest birefringence of the common minerals, and its high order pink interference color is so pale that it is commonly mistaken for a first order white. Examination of the edges of crystals will, however, reveal multiple isochromatic lines. Rhombohedral cleavage and twinning are very common. Crossed polars, X35.

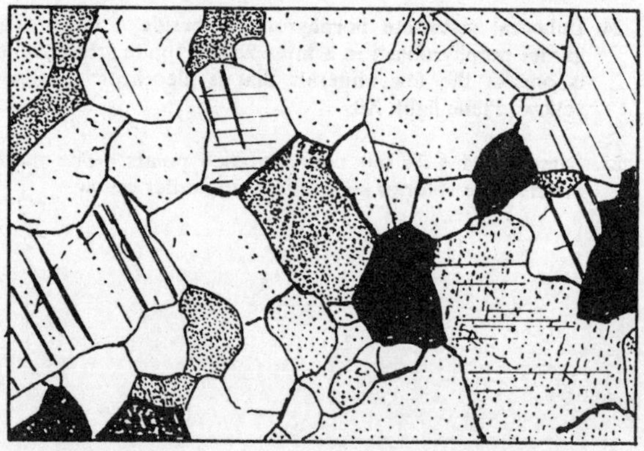

33. Prisms of tremolite and anhedral diopside with quartz produced from low-grade metamorphism of impure dolomitic limestone. Plane light, X20.

34. Same as MM 33 but under crossed polars. Tremolite shows up to second order blue interference colors.

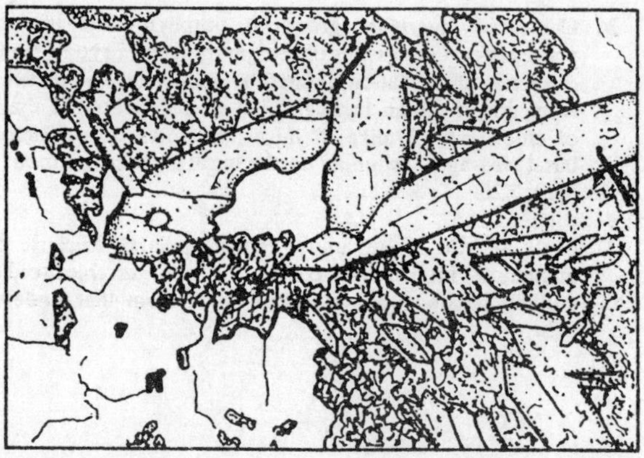

METAMORPHIC ROCK-FORMING MINERALS

35. Partially serpentinized forsteritic olivine in marble. Crossed polars, X20.

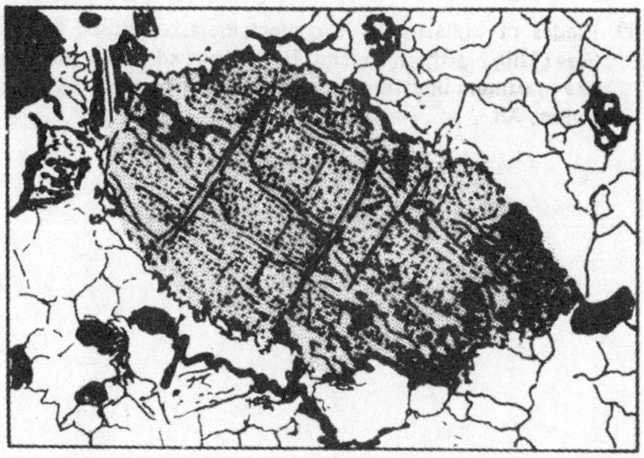

36. Forsterite in marbles is commonly completely altered to serpentine, forming a rock known as ophicalcite. Crossed polars, X8.

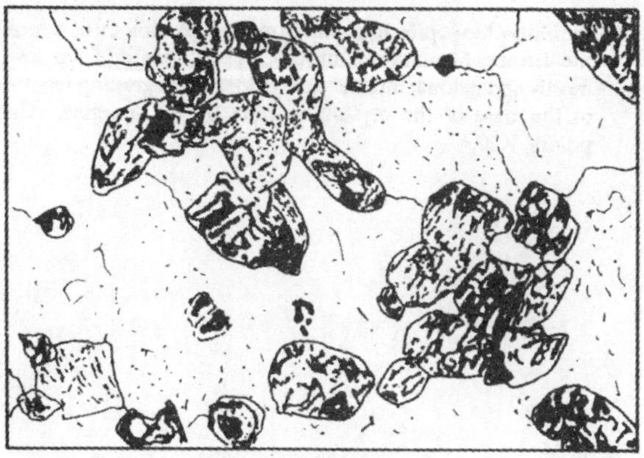

37. Diopside- and scapolite-bearing marble. Under crossed polars diopside and scapolite exhibit similar middle second order maximum interference colors. Under plane light, however, diopside has strong positive relief (grain at middle left), whereas scapolite has very low relief (upper right) and resembles plagioclase. Scapolite is one of the few uniaxial negative minerals to exhibit moderate birefringence. Crossed polars, X20.

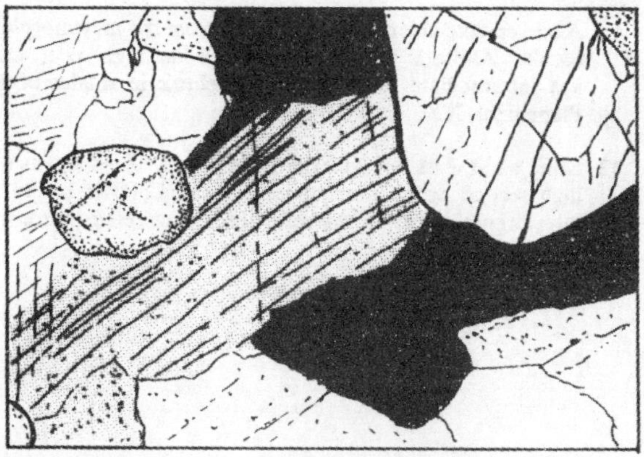

38. Vesuvianite (idocrase) intergrown with diopside and quartz developed by contact metamorphism of impure limestone. Vesuvianite has very low birefringence and exhibits anomalous blue interference colors. Crossed polars, X80.

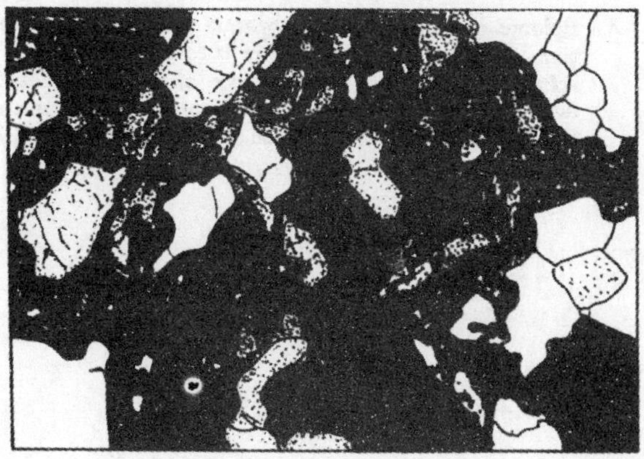

METAMORPHIC ROCK-FORMING MINERALS

39. Blades of wollastonite in contact metamorphosed limestone, Scawt Hill, Northern Ireland. Under crossed polars wollastonite has maximum interference color of first order yellow. Crossed polars, X8.

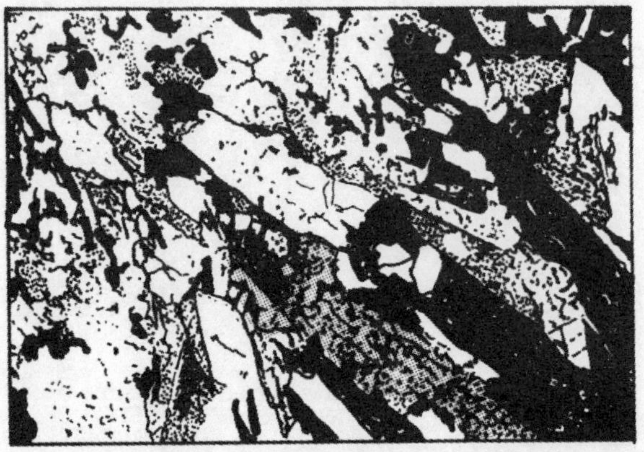

40. Melilite-phlogopite hornfels from impure dolomitic xenolith in the Brome Mountain gabbro, Quebec. Melilite has low, and mostly anomalous, interference colors, here grading from brown in the core of the crystals to blue on the margins. Crossed polars, X20.

41. Albite-epidote-chlorite schist formed from the metamorphism of basalt. Albite is untwinned; epidote has high relief; chlorite has intermediate relief and is pleochroic in shades of green. Plane light, X20.

42. Same as MM 41 but under crossed polars. Epidote has bright, high second order interference colors. Chlorite has very low interference colors, in this case, anomalous reddish brown.

43. Epidote in contact with an opaque grain grades outward to clinozoisite at the margin where it contacts plagioclase. The refractive indices and birefringence of epidote decrease with decreasing iron content; clinozoisite, the iron-poor epidote, has anomalous blue interference colors. Crossed polars, X10.

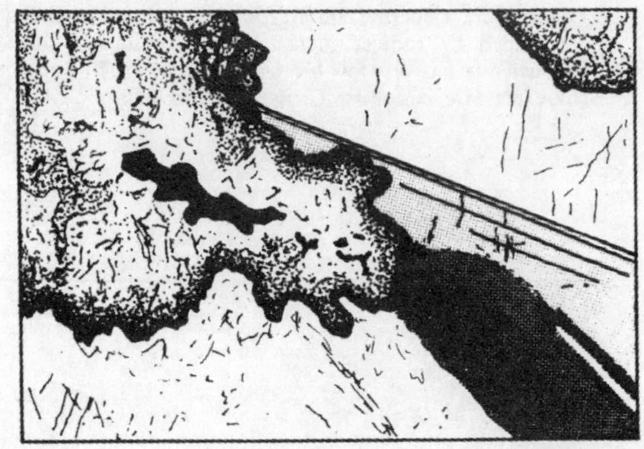

METAMORPHIC ROCK-FORMING MINERALS

44. Piemontite, the manganese epidote, is pleochroic from red to canary yellow; the orientation in this photograph gives red. In contact with phlogopite and quartz. Plane light, X50.

45. Same as MM 44 but polarizer has been rotated through 90° so that piemontite is now yellow.

46. Actinolite-chlorite-epidote schist (meta-basalt). Chlorite has low relief relative to the other minerals. Plane light, X25.

47. Same as MM 46 but under crossed polars where epidote is seen to have much higher birefringence than actinolite, and chlorite has anomalous bluish gray interference colors.

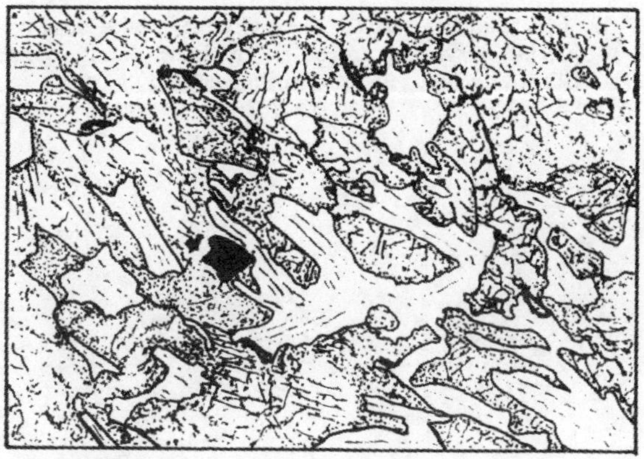

48. Amphibolite containing colorless cummingtonite, green hornblende, and untwinned plagioclase. Both amphiboles contain thin basal exsolution lamellae of the other. Where crystals are twinned, the exsolution lamellae produce a herringbone pattern. Plane light, X25.

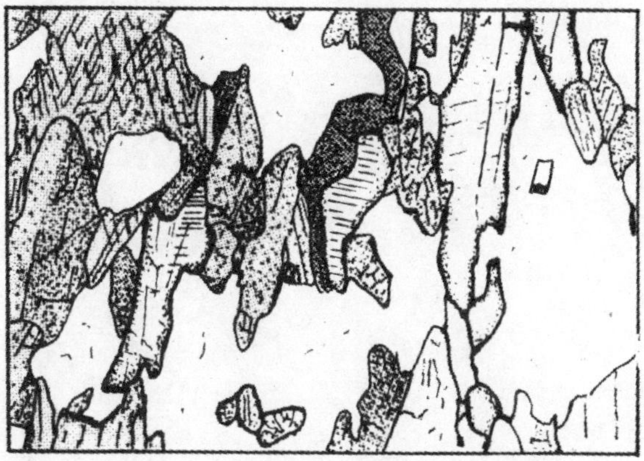

49. Glaucophane schist showing the characteristic blue pleochroism of this amphibole. Plane light, X20.

METAMORPHIC ROCK-FORMING MINERALS

50. Lawsonite crystals in glaucophane schist. Lawsonite is colorless. Plane light, X20.

51. Same as MM 50 but under crossed polars where lawsonite is seen to have bright interference colors like those of epidote.

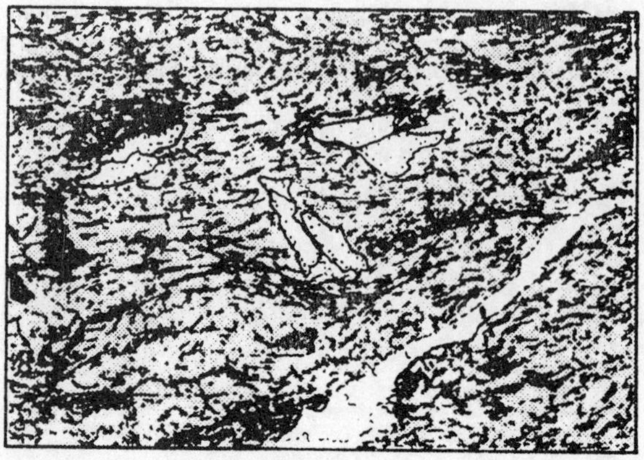

52. Jadeite quartz schist. The pyroxene jadeite does not form distinct crystals; instead it typically forms bundles of slightly diverging fibers. Plane light, X20.

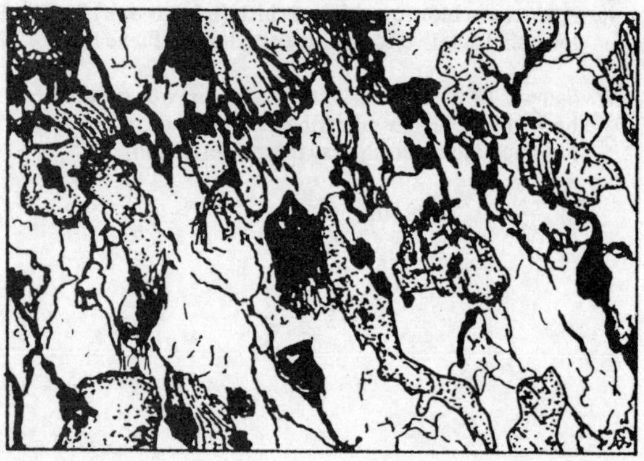

4 Classification of Igneous Rocks

Igneous rocks typically consist of seven major oxides, SiO_2, Al_2O_3, FeO (Fe_2O_3), MgO, CaO, Na_2O, and K_2O, which combine to form the *rock-forming minerals*, which include clinopyroxene, orthopyroxene, olivine, garnet (only at high pressure), amphibole, mica, quartz, plagioclase, alkali feldspar, feldspathoid, and spinel (commonly magnetite). Most of the minerals belong to solid solution series, and thus they can adjust their compositions to fit a wide range of bulk rock compositions. As a result, most rocks are composed of no more than four of the rock-forming minerals, and some contain less.

Classifications of igneous rocks have been based on both the seven major oxides and the eleven essential rock-forming minerals. A chemical classification is of little use in the field, because it requires analytical data. Minerals, on the other hand, are readily identifiable in the field, if sufficiently coarse grained. They may also provide valuable information about the environment in which a rock forms, which the chemical classification does not. Coesite and quartz, for example, are chemically identical, but coesite forms only under extremely high pressures. A mineralogical classification, however, is of little use if the minerals are too fine-grained to be identifiable or the rock is glassy; in these cases the chemical classification would be more useful.

This laboratory manual uses the classification proposed by the International Union of Geological Sciences (IUGS) Subcommission on the Systematics of Igneous Rocks (Streckeisen, 1976, 1979; LeBas et al., 1986). For volcanic rocks, the Irvine and Baragar (1971) classification is also included. The IUGS classification uses both the mineralogical and chemical classifications, but emphasizes the mineralogical one wherever possible.

Mode and Norm

The mineralogical composition of a rock is known as its *mode*. The abundances of the minerals are normally expressed in terms of volume percent, for this is what the eye actually perceives. Accurate modes are obtained by *point counting* thin sections of rock under the petrographic microscope. This method involves identifying the mineral beneath the crosshairs of the microscope each time the sample is advanced a metered amount by a mechanical stage. The metered increments essentially create a grid of points on the sample, and the fraction of the total number of points occupied by a particular mineral is proportional to the volume percent of that mineral in the rock. An approximate mode can be obtained by simply estimating the percentages of the minerals in the hand specimen or thin section. The figures on the inside back cover, which give examples of a microscope field of view filled with various percentages of a phase, will help make this task easier. Be careful when determining modes that a low enough magnification is used to sample a representative area of the rock. With extremely coarse-grained rocks, an entire thin section may not even be representative, and several sections may need to be measured. Finally, if the rock is layered, the section must cut across, and preferably be normal to, the layering or lineation if a representative mode is to be obtained.

If a rock is too fine-grained for the mode to be determined under the microscope, or if it consists of a substantial amount of glass, a chemical analysis will be needed to classify the rock. Modern automated x-ray fluorescence spectrometers can produce a major element analysis of a rock in less time than it takes to do a normal thin section study of a rock. Indeed, if such equipment is fitted with an energy dispersive detector (a solid state device that detects the energy of x-rays), a few minutes are all that are necessary for an analysis. This does not include sample preparation, which takes about the same length of time as the preparation of a thin section.

In order to classify a rock on the basis of a chemical analysis, it is necessary to convert the amounts of the various oxides into the minerals that would have formed if the magma had cooled slowly. The calculated abundance of the minerals is known as a *norm*. The oxides in the rock analysis are allocated, following a prescribed set of rules, to simple anhydrous end-member formulae of the rock-forming and common accessory minerals. The normative minerals are the typical ones that would form if a magma crystallized at low pressure, for example in a lava flow. The normative composition of a rock formed at low pressure will therefore be similar to its mode. You may wish to check this by point counting, for example, a coarse-grained analyzed basalt or granite. (For high pressures or for hydrous conditions special normative calculations have been made.) The norm, then, allows a chemical analysis of a rock to be recast in terms of the common minerals which are the basis of the IUGS classification.

The first norm was devised by the four petrologists Cross, Iddings, Pirsson, and Washington (Johannsen, 1931), and thus it is commonly referred to as the **CIPW norm.** It takes the weight percent of the oxides and recasts them as the weight percent of the minerals; it is therefore also known as a *weight norm*. The normative percentages are not directly comparable to those of the mode, which are listed as volume percent. The discrepancy is particularly noticeable for very dense minerals such as magnetite and ilmenite. The weight norm, however, can be converted to volume proportions by dividing the weight percentage of each normative mineral by its density and recalculating to percentages. Normative calculations have also been presented in terms of molecular proportions; these have the advantage of deviating less from the volume mode. Because all such calculations are now carried out on computers, there is no reason not to convert the normative mineral percentages to volume percent if comparison with a mode is to be made. Remember that the IUGS classification is based on volume percentages.

The task of distributing the thirteen oxides normally given in a rock analysis amongst a multitude of common minerals at first seems impossible. For example, where should CaO be allotted? Anorthite, clinopyroxene, and wollastonite all contain CaO, as do the accessory minerals apatite and calcite. Silica is still more of a problem because most of the common minerals are silicates, but they do not all require the same amounts of silica. Of course, there would be no unique solution to this problem if there were no restrictions on the minerals that form. Petrographic experience, however, indicates that nature places severe restrictions on the possible mineral assemblages that can occur. On the basis of these restrictions, then, Cross, Iddings, Pirsson, and Washington devised their scheme for calculating the norm. It is a credit to their skills as petrographers that the norm calculation has stood the test of time and can now even be justified thermodynamically. Although the calculation need not be memorized, valuable petrographic insight is gained from a general appreciation of the steps involved, in particular with those dealing with the distribution of silica.

CIPW Norm Calculation

To begin the CIPW norm calculation, the weight percentages of the oxides in the rock analysis are converted to mole proportions by dividing each oxide weight percent by its molecular weight. Because trace amounts of Mn and Ni substitute for Fe in the common ferromagnesian minerals, the mole proportions of MnO and NiO are added to that of FeO; similarly the mole proportions of BaO and SrO are added to that of CaO. Using the formulae given in Table 4-1, the mole proportions are distributed amongst the normative minerals according to the rules given below. The calculations can be carried out conveniently on a balance sheet, as shown in Table 4-2. The first column gives the chemical analysis of the rock in weight percent; this is followed by the molecular weights of the oxides; the third column gives the molecular proportions. The remaining columns, one for each normative mineral formed, show the amount of an oxide that remains after the necessary amount has been allocated to the particular mineral. Eventually, each of the oxides is reduced to zero at the right side of the table after all of the oxides have been allotted to the appropriate minerals. The weight percentages of the normative minerals are then calculated by multiplying the appropriate oxide by the factor given in Table 4-1. As a check on the calculation, the sum of the normative minerals should be the same as the sum of the weight percentages of the oxides used in the original analysis (H_2O is ignored).

In the following rules, the name of an oxide refers to its mole proportion. Thus, a rule stating that the SiO_2 is to be reduced by 0.5 x MgO indicates that the mole proportion of SiO_2 is to be reduced by an amount equal to one half the mole proportion of MgO.

1) P_2O_5 is allotted to apatite, and CaO is reduced by 3.33 x P_2O_5.
2) S is allotted to pyrite, and FeO is reduced by 0.5 x S.
3) Cr_2O_3 is allotted to chromite, and FeO is reduced by Cr_2O_3.
4) TiO_2 is allotted to ilmenite, and FeO is reduced by TiO_2. If TiO_2 exceeds FeO, the excess is allotted to provisional sphene (tn'), and CaO and SiO_2 are both reduced by an amount equal to the excess of TiO_2; this step is carried out only if CaO remains after formation of anorthite (step 10). If TiO_2 still remains, it is calculated as rutile.
5) F is allotted to fluorite, and CaO is reduced by 0.5 x F.
6) CO_2 is allotted to calcite, and CaO is reduced by CO_2.
7) ZrO_2 is allotted to zircon, and SiO_2 is reduced by ZrO_2.
8) K_2O is allotted to provisional orthoclase (or'), and Al_2O_3 is reduced by K_2O, and SiO_2 is reduced by 6 x K_2O.
9) Al_2O_3 remaining from step 8 is combined with an equal amount of Na_2O to form provisional albite (ab'), and silica is decreased by 6 times this amount. If there is insufficient Al_2O_3, proceed to step 11.
10) Al_2O_3 remaining from step 9 is combined with an equal amount of CaO to form provisional anorthite (an'), and silica is decreased by twice this amount. If Al_2O_3 exceeds CaO, it is calculated as corundum.
11) If Na_2O exceeds Al_2O_3 in step 9, an amount of Fe_2O_3 equal to the excess is allotted to acmite, and silica is decreased by 4 times this amount.
12) If Na_2O still remains after step 11, the remaining Na_2O forms sodium metasilicate, and silica is reduced by the amount of the remaining Na_2O (this is extremely rare).

TABLE 4-1 Normative Minerals

Mineral	Abbrev.	Formula	Mol. to Wt. Factor		
Quartz	Q	SiO_2	SiO_2	x	60.09
Corundum	C	Al_2O_3	Al_2O_3	x	101.96
Zircon	Z	$ZrO_2.SiO_2$	ZrO_2	x	183.31
Orthoclase	or	$K_2O.Al_2O_3.6SiO_2$	K_2O	x	556.67
Albite	ab	$Na_2O.Al_2O_3.6SiO_2$	Na_2O	x	524.46
Anorthite	an	$CaO.Al_2O_3.2SiO_2$	CaO	x	278.21
Leucite	lc	$K_2O.Al_2O_3.4SiO_2$	K_2O	x	436.50
Nepheline	ne	$Na_2O.Al_2O_3.2SiO_2$	Na_2O	x	284.11
Kaliophilite	kp	$K_2O.Al_2O_3.2SiO_2$	K_2O	x	316.33
Acmite	ac	$Na_2O.Fe_2O_3.4SiO_2$	Na_2O	x	462.02
Na metasilicate	ns	$Na_2O.SiO_2$	Na_2O	x	122.07
	wo	$CaO.SiO_2$	CaO	x	116.17
Diopside	en	$MgO.SiO_2$	MgO	x	100.39
	fs	$FeO.SiO_2$	FeO	x	131.93
Wollastonite	wo	$CaO.SiO_2$	CaO	x	116.17
Hypersthene	en	$MgO.SiO_2$	MgO	x	100.39
	fs	$FeO.SiO_2$	FeO	x	131.93
Olivine	fo	$2MgO.SiO_2$	MgO	x	70.35
	fa	$2FeO.SiO_2$	FeO	x	101.89
Ca orthosilicate	cs	$2CaO.SiO_2$	CaO	x	86.12
Magnetite	mt	$FeO.Fe_2O_3$	FeO	x	231.54
Chromite	cm	$FeO.Cr_2O_3$	Cr_2O_3	x	223.84
Hematite	hm	Fe_2O_3	Fe_2O_3	x	159.69
Ilmenite	il	$FeO.TiO_2$	TiO_2	x	151.75
Sphene (titanite)	tn	$CaO.TiO_2.SiO_2$	TiO_2	x	196.07
Perovskite	pf	$CaO.TiO_2$	TiO_2	x	135.98
Rutile	ru	TiO_2	TiO_2	x	79.90
Apatite	ap	$3(3CaO.P_2O_5).CaF_2$	P_2O_5	x	336.21
Fluorite	fr	CaF_2	F	x	39.04
Pyrite	pr	FeS_2	S	x	59.98
Calcite	cc	$CaO.CO_2$	CO_2	x	100.09

13) All remaining Fe_2O_3 is allotted to magnetite, and the FeO is decreased by Fe_2O_3. If Fe_2O_3 exceeds FeO, the excess is calculated as hematite.

14) All remaining MgO and FeO forms pyroxenes and olivines. At this point, MgO and FeO are added together, but their proportions are maintained in calculating the amounts of the Mg and Fe end-member-components of the pyroxenes and olivines.

15) The CaO remaining from step 10 forms provisional diopside (di'), which decreases MgO + FeO by an amount equal to CaO, and silica by twice this amount.

16) If CaO exceeds MgO + FeO, the excess forms provisional wollastonite (wo'), and silica is decreased by the excess CaO.

17) If the MgO + FeO in step 15 exceeds the CaO, the excess forms provisional hypersthene, and silica is decreased by the excess MgO + FeO.

18) If SiO_2 is still positive, remaining SiO_2 is calculated as quartz.

19) If SiO_2 is negative, the rock contains insufficient silica for the provisionally formed silicate minerals. Some of these minerals must therefore be converted to ones containing less silica, until the silica deficiency is eliminated. The order in which this is done is as follows: first, hypersthene is converted to olivine, then sphene to perovskite, albite to nepheline, orthoclase to leucite, wollastonite and diopside to calcium orthosilicate and olivine, and finally leucite to kaliophilite. Let the deficiency in silica be D; the amounts of the provisional normative minerals are designated with a prime (see Table 4-1 for abbreviations).

20) If D < hy/2', set ol = D, and hy = hy' - 2D.
If D > hy'/2, all provisional hypersthene is converted to olivine (ol = hy'), and the new silica deficiency, D_1, is D - hy'/2.

21) If D < tn', set pf = D_1 and tn = tn' - D_1.
If D > tn', all provisional sphene is converted to perovskite (pf = tn'), and the new silica deficiency, D_2, is D_1 - tn'.
If no provisional sphene was made in step 4, simply set $D_2 = D_1$.

22) If D_2 < 4ab', some of the provisional albite is converted to nepheline, such that ne = $D_2/4$ and ab = ab' - $D_2/4$.
If D_2 > 4ab', all provisional albite is converted to nepheline (ne = ab'), and the new silica deficiency, D_3, is D_2 - 4ab'.

23) If D_3 < 2or', some provisional orthoclase is converted to leucite, such that lc = $D_3/2$ and or = or' - $D_3/2$.
If D_3 > 2or', all provisional orthoclase is converted to provisional leucite (lc' = or') and the new silica deficiency, D_4, is D_3 - 2or'.

24) If D_4 < wo'/2 in provisional wollastonite (not in diopside), some provisional wollastonite is converted to calcium orthosilicate, such that cs = D_4, and wo = wo' - $2D_4$.
If D_4 > wo'/2, all provisional wollastonite is converted to calcium orthosilicate (cs = wo'/2), and the new silica deficiency, D_5, is D_4 - wo'/2.

25) If D_5 < di', some of the diopside is converted to calcium orthosilicate and olivine which are added to that previously formed; set cs = $D_5/2$, ol = $D_5/2$, and di = di' - D_5, remembering to add the amounts of cs and ol to those already formed in steps 24 and 20, respectively.
If D_5 > di', all provisional diopside is converted to calcium orthosilicate and olivine, such that cs = di'/2 and ol = di'/2 (add to amounts formed in steps 24 and 20), and the new silica deficiency, D_6, is D_5 - di'.

26) Finally, if there is still a deficiency in silica, leucite is converted to kaliophilite; set kp = $D_6/2$, and lc = lc' - $D_6/2$.

Once the silica deficiency has been eliminated (one of the steps between 20 and 26), the norm calculation is completed by multiplying the mole proportion of the first oxide in the formula of each normative mineral formed by the weight-conversion factor given in Table 4-1. Only rare rock types have low enough silica contents to cause the norm calculation to proceed beyond step 22. For most rocks, the calculation is therefore quite short. Nonetheless, if large numbers of norms have to be determined, the calculations become tedious, and they are preferably done using a computer. Those students interested in computer programming or who are familiar with computer spread sheets will find the construction of a norm program an interesting and useful exercise.

TABLE 4-2. Example of Norm Calculation

				If Mole Proportion is affected by the formation of the normative mineral, the new value is entered here.											
										di			hy		
Oxide	Wt%	Mol Wt	Mol Prop	ap	il	or	ab	an	mt	wo	en	fs	en	fs	Q
SiO_2	53.80	60.09	.8953	----	----	.7998	.5094	.3654		.3059	.2745	.2464	.1761	.1132	.0000
TiO_2	2.00	79.9	.0250	----	.0000										
Al_2O_3	13.90	101.96	.1363	----	----	.1204	.0720	.0000							
Fe_2O_3	2.60	159.69	.0163	----	----	----	----	----	.0000						
FeO	9.30	71.85	.1294	----	.1072	----	----	----	.0909	----	----	.0629	----	.0000	
MnO	0.20	70.94	.0028												
MgO	4.10	40.31	.1017	----	----	----	----	----	----	----	.0703	----	.0000		
CaO	7.90	56.08	.1409	.1315	----	----	----	.0595		.0000					
Na_2O	3.00	61.98	.0484	----	----	----	.0000								
K_2O	1.50	94.2	.0159	----	----	.0000									
P_2O_5	0.40	141.95	.0028	.0000											
Total	98.70														
Mole Propor. of Normative Mineral				.0028	.0250	.0159	.0484	.0720	.0163	.0595	.0314	.0281	.0703	.0629	.1132
Mol. to Wt Conversion Factor				336	152	557	524	278	232	116	100	132	100	132	60.1
Wt% of Normative Minerals				0.95	3.80	8.86	25.39	20.03	3.77	6.91	3.15	3.70	7.06	8.29	6.80
Sum of Normative Minerals				98.72											

Calculation of diopside		
MgO + FeO + MnO 0.1927		In diopside
MgO/(FeO + MnO) 1.1184	= en/fs	Wo = .0595
		En = .0314
		Fs = .0281

General Classification Terms

Rocks are divided into *plutonic, hypabyssal,* and *extrusive* or *volcanic* depending on whether they are emplaced at great depth, near the surface, or on the surface, respectively. No definite division separates plutonic from hypabyssal rocks; in general the distinction is based on grain size. In plutonic rocks, minerals are readily identifiable in hand specimen because grain sizes are either medium or coarse (>1 mm). Hypabyssal and volcanic rocks are fine-grained or even glassy, and mineral identification in hand specimen is difficult. Hypabyssal and volcanic rocks are therefore commonly classified together.

The general composition of a rock can be indicated by a number of different but equivalent terms (Table 4-3). Rocks rich in quartz, feldspars, or feldspathoids are said to be *felsic*, whereas those rich in ferromagnesian minerals are *mafic* or even *ultramafic* if they are totally devoid of felsic minerals. Felsic rocks are rich in SiO_2, whereas ultramafic rocks are rich in MgO and FeO. Early in the development of petrology, silica in magmas was thought to combine with water to form siliceous acid, whereas MgO and FeO were thought to form bases. This led to the terms *acid, basic*, and *ultrabasic*, which although obsolete are still commonly used. Application of any of these terms in the field is based on the simple observation of the color of the hand specimen, and thus it seems more honest to use terms that indicate this. Light-colored rocks are referred to as *leucocratic*, whereas dark ones are *melanocratic*. A quantitative statement of the color of a rock can be made by using the *color index*, which is simply the volume percentage of the dark ferromagnesian minerals. The color index can also be based on the percentage of ferromagnesian minerals in the CIPW norm if an accurate mode is not available.

TABLE 4-3 General Descriptive Petrographic Terms

Mineralogical		Chemical		Obsolete		Color		Color index
Felsic	=	SiO_2-rich	=	Acid	=	Leucocratic	=	0
Mafic	=		=	Basic				|
Ultramafic	=	(Mg,Fe)O-rich	=	Ultrabasic	=	Melanocratic	=	100

IUGS Classification of Plutonic Igneous Rocks

The classification of plutonic rocks is less controversial than that of volcanic ones, because the constituent minerals are readily identifiable and thus can be used for modal classification. The following is a summary of the classification of plutonic rocks recommended by the IUGS Subcommission on the Systematics of Igneous Rocks (Streckeisen, 1976).

The IUGS classification is based on the modal amounts of the common minerals, which are divided into five groups:

- **Q** quartz
- **A** alkali feldspar, including albite with up to 5 mol% anorthite ($<An_5$)
- **P** plagiocalse with composition from An_5 to An_{100} and scapolite (a common alteration product of plagioclase)
- **F** feldspathoids: nepheline, sodalite, analcite, leucite, pseudoleucite, kalsilite (kaliophilite), nosean, hauyne, and cancrinite
- **M** mafic minerals: olivine, pyroxenes, amphiboles, micas, monticellite, melilite, opaque minerals, and accessory minerals, such as zircon, apatite, sphene, epidote, allanite, garnet, and carbonate.

Rocks containing less than 90% mafic minerals ($M<90$) are classified on the basis of their proportions of Q, A, P, and F; rocks with $M>90$ are classified on the basis of the proportions of the major mafic minerals. The division at $M=90$ is arbitrary, but few rocks have compositions near this division. Ultramafic rocks typically have $M>>90$, whereas other rocks fall well below this limit.

Because rocks never contain both Q and F, nonultramafic rocks can be classified in terms of three components, either QAP or FAP. The three components can be represented either in triangular plots or in the orthogonal plot of Fig. 4-1. The horizontal axis in this diagram indicates the proportion of plagioclase in the total feldspar ($P/\{P+A\}$), and the vertical axis indicates the modal percentage of quartz (Q) measured upward and the amount of feldspathoid (F) measured downward from the zero line. The diagram is subdivided into areas marking the compositional extent of the main rock types. Although the divisions are arbitrary, they are positioned so that rocks that most petrologists associate with a particular name occupy the center of a field. Thus a rock containing equal amounts of alkali feldspar and plagioclase and 10% quartz would be named a quartz monzonite; a rock containing 15% nepheline and having feldspar that is only 5% plagioclase would be named a nepheline syenite.

Some rocks cannot be uniquely defined in terms of only QAP or FAP, and further diagnostic properties must be used. For example, as plagioclase becomes more abundant to the right in Fig. 4-1, it also becomes more anorthite-rich. But in some rocks it becomes more anorthite-rich than in others. This difference serves to distinguish *diorite* from *gabbro*; the average plagioclase composition in diorite is less than An_{50}, whereas in gabbro it is greater than An_{50}. Zoning of plagioclase crystals may make this distinction difficult to apply.

Rocks generally become more mafic with increase in plagioclase content, as indicated in the graphs at the top and bottom of Fig. 4-1. Rocks rich in quartz have the lowest contents of mafic minerals, whereas those containing feldspathoids tend to have the highest. Also, gabbroic rocks are generally more mafic than dioritic ones. For any particular rock type there is a normal range of concentration of mafic minerals, as shown in Fig. 4-1. Rocks containing more or less than this can have the prefix *mela-* or *leuco-*, respectively, added to the name. Thus, a gabbro containing a large amount of plagioclase would be described as a leucogabbro. If the amount of plagioclase is exceptionally high (>90) the rock is given the special name *anorthosite*. Further definition of a rock type can be given by specifying the names of the major mafic minerals present, with the most abundant being placed closest to the rock name. Thus, a biotite-hornblende diorite contains more hornblende than biotite.

Although gabbroic rocks plot in Fig. 4-1, their compositional variations cannot be expressed in terms of the components of this diagram. Gabbroic rocks consist essentially of plagioclase, orthopyroxene, clinopyroxene, olivine, and hornblende, which can be used to classify them into a number of major types (Fig. 4-2). Normal gabbroic rocks contain between 35 and 65% mafic minerals; if they contain more or less than this they are prefixed by the terms mela- or leuco- respectively. In addition to the major components, gabbroic rocks may also contain magnetite, ilmenite, spinel, biotite, and garnet, which can be added as qualifiers to the rock name, again with the most abundant mineral being placed closest to the rock name. The plagioclase in gabbroic rocks has an average composition $>An_{50}$, but more albitic compositions may occur on the rims of zoned crystals. Some anorthosites are an exception to this rule. Although anorthosites in layered gabbroic intrusions contain plagioclase in the bytownite range, large Precambrian anorthosite massifs may consist entirely of andesine plagioclase (some are composed of labradorite). Nonetheless, massif type anorthosites have been classified with gabbros rather than with diorites, but many of their associated rocks bear more affinities to the diorite-granodiorite suite of rocks than to the gabbroic suite.

Ultramafic rocks are composed essentially of olivine, orthopyroxene, clinopyroxene, and hornblende. The names used for the various types of ultramafic rock are shown in Fig. 4-3. Small amounts of spinel, garnet, biotite, magnetite, or chromite can be indicated, for example, as spinel-bearing peridotite if the amount is $\leq$ 5% and spinel peridotite if >5%. If the amount exceeds 50%, as might happen, for example, with chromite, the rock is referred to as a peridotitic chromitite, and if the amount exceeds 95% it is named a *chromitite*. One unusual type of phlogopite-rich peridotite that contains mantle-derived xenoliths is known as *kimberlite* because of its occurrence at Kimberly, South Africa, where it forms the host rock for diamonds.

Many igneous rocks, especially the more mafic feldspathoid-bearing ones, contain carbonate which has formed as a late-crystallizing phase or as a late alteration product, but it rarely exceeds a few percent. A group of igneous rocks known collectively as *carbonatites*, however, are composed essentially of carbonates (>50% carbonate). These have such low silica contents that they cannot be classified by the schemes for silicate rocks. Most are composed of mixtures of calcite and dolomite, but iron carbonate may be present as well. Recent eruptions of Oldoinyo Lengai

Figure 4-1 IUGS classification of igneous rocks containing less than 90% mafic minerals. Plutonic rocks are in upper case and volcanic ones in lower case letters. A = alkali feldspar + albite ($<An_5$) and P = plagioclase ($>An_5$).

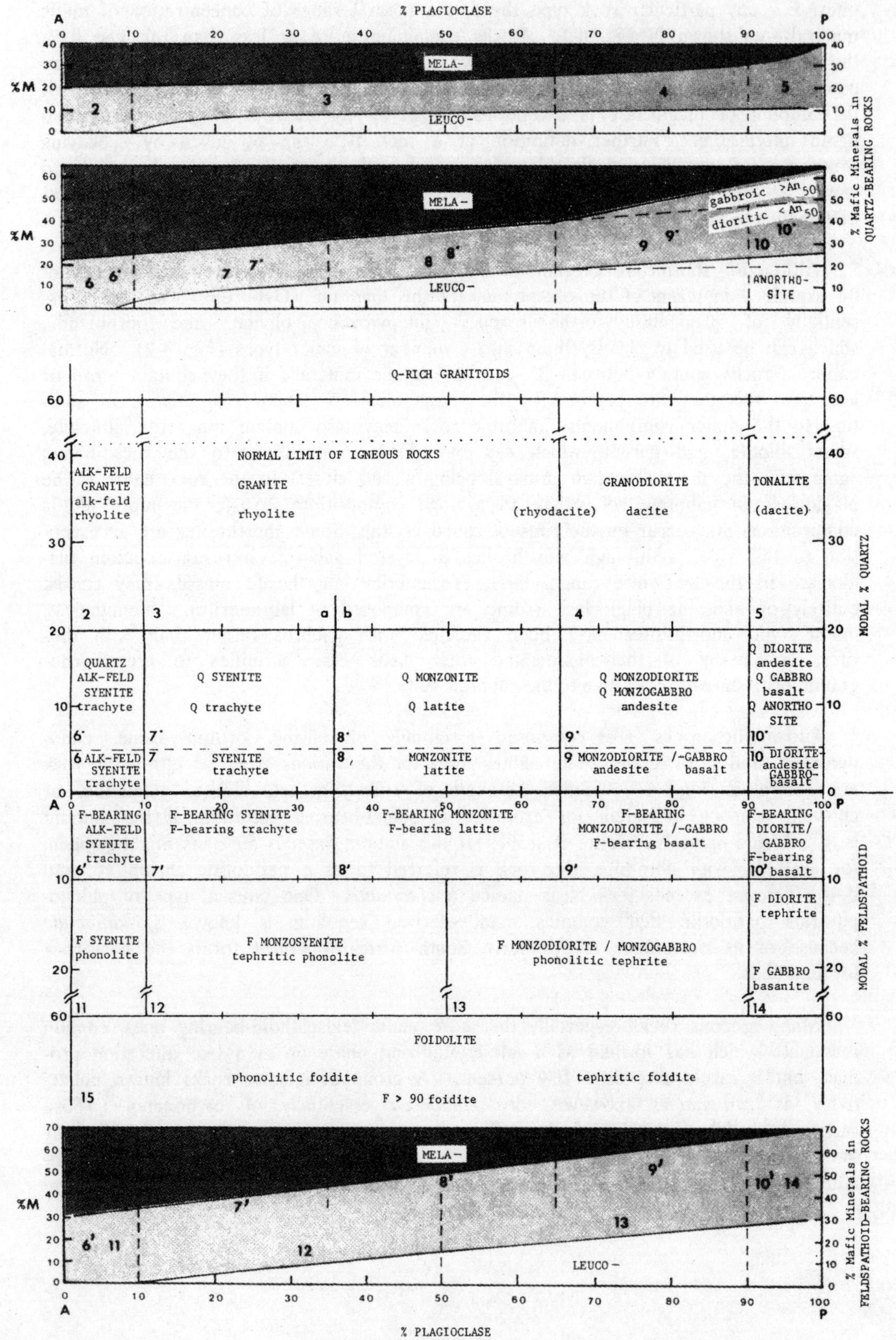

TABLE 4 - 4 Common rock names, with IUGS-recommended terms in bold print (The first number refers to either Fig. 4-1, 4-2, or 4-3; numbers following the colon give the compositional field in which the rock plots in the figures.)

Adamellite	1:8*	plutonic
Alaskite	1:2, 1	plutonic, no mafics
Alkali feldspar		
" **granite**	1:2	plutonic, name feldspar
" **rhyolite**	1:2	volcanic, name feldspar
" **syenite**	1:6	plutonic, name feldspar
" **trachyte**	1:6	volcanic, name feldspar
Alkali granite	1:2,3	plutonic, alkali amph or pyrox
Alkali rhyolite	1:2,3	volcanic, alkali amph or pyrox
Alkali syenite	1:6*,7*	plutonic, alkali amph or pyrox
Alkali trachyte	1:6*,7*	volcanic, alkali amph or pyrox
Alnoite	Table 4-5	hypabyssal
Andesite	1:10,10*,9,9*	volcanic
Ankaramite	1:10	volcanic, porph., ol-aug-rich
Anorthosite	2:1	plutonic cumulate
Aplite	1:3	suggary texture
Basalt	1:10*,10,10'	volcanic
Basanite	1:14	volcanic
Benmorite	1:8,8'	volcanic
Camptonite	Table 4-5	hypabyssal
Carbonatite	see text	plutonic, rare volc. carbonate
Charnockite	1:3b	plutonic, hypersthene
Commendite	1:2	volcanic, peralkaline
Cortlandite	2:6	plutonic, poikilitic texture
Dacite	1:4	volcanic
Diabase	1:10,10*	hypabyssal
Diorite	1:10,10*	plagioclase $<An_{50}$
Dolerite	1:10,10*	hypabyssal
Dunite	3:1	plutonic, cumulate
Essexite	1:10'	plutonic
Fenite	1:7	aegerine, metasomatic
Foid syenite	1:11	plutonic, name feldspathoid
Fourchite	Table 4-5	monchiquite but no olivine
Foyaite	1:11	plutonic
Gabbro	1:10	plutonic
Granite	1:3	plutonic
Granodiorite	1:4	plutonic
Granophyre	1:3	hypabyssal, micrographic text
Harzburgite	3:4	plutonic, cumulate
Hornblendite	3:9	plutonic
Icelandite	1:9*	volcanic, Fe-rich
Ijolite	1:15	nepheline--aegerine, plutonic
Jacupirangite	1:15	titanaug--nepheline, plutonic
Jotunite	1:9	plutonic, hypersthene
Keratophyre	1:8*	meta-volcanic, albite
Kersantite	Table 4-5	hypabyssal
Kimberlite	-	ultramafic, phlog, hypabyssal
Komatiite	3:3	volcanic
Lamproite	Table 4-5	similar to lamprophyre, volcanic, K_2O and MgO-rich
Larvikite	1:6,6*	plutonic
Latite	1:8	volcanic
Lherzolite	3:3	plutonic, cumulate

Mangerite	1:8	plutonic, hypersthene, mesoperthite
Melilitite	-	melilite, volcanic
Minette	Table 4-5	hypabyssal
Monchiquite	Table 4-5	hypabyssal
Monzonite	1:8	plutonic
Monzodiorite	1:9	plutonic
Mugearite	1:9,9'	volcanic
Neph. syenite	1:11	plutonic
Nephelinite	1:15	volcanic
Nordmarkite	1:7,7*	plutonic
Norite	2:3	plutonic
Pantellerite	1:,2,6*	volcanic, peralkaline
Peridotite	3:2,3,4	plutonic, cumulate
Phonolite	1:11	volcanic
Picrite	1:10	volcanic, porph. ol-rich
Polzenite	Table 4-5	hypabyssal
Pulaskite	1:6'	plutonic
Pyroxenite	3:5,6,7	plutonic, cumulate
Qtz. diorite	1:10*	plutonic
Qtz. latite	1:8*	volcanic
Qtz.keratophyre	1:3	meta-volcanic, albite
Qtz. monzonite	1:8*	plutonic
Qtz. syenite	1:7*	plutonic
Qtz. trachyte	1:7*	volcanic
Rapakivi	1:3	plutonic, plagioclase rimming K-feld
Rhyodacite	1:3-4	volcanic
Rhyolite	1:3	volcanic
Sannaite	Table 4-5	hypabyssal
Shonkinite	1:7'	plutonic, hypabyssal, mafic
Spessartite	Table 4-5	hypabyssal
Spilite	1:10	meta-basalt, albite
Syenite	1:7	plutonic
Syenodiorite	1:9	plutonic
Tephrite	1:14	volcanic, plag $< An_{50}$
Teschenite	1:14	hypabyssal
Theralite	1:14	plutonic
Tholeiite	1:10*	volcanic
Tinguaite	1:11	hypabyssal, analcite
Tonalite	1:5	plutonic
Trachyte	1:7	volcanic
Tristanite	1:8'	volcanic
Troctolite	2:5	plutonic, plag cumulate
Trondhjemite	1:4	plutonic, sodic
Ugandite	1:15	volcanic, mafic, leucite
Vogesite	Table 4-5	hypabyssal
Websterite	3:6	plutonic, cumulate
Whehrlite	3:2	plutonic, cumulate
Wyomingite	1:15	volcanic, phlogopite, leucite

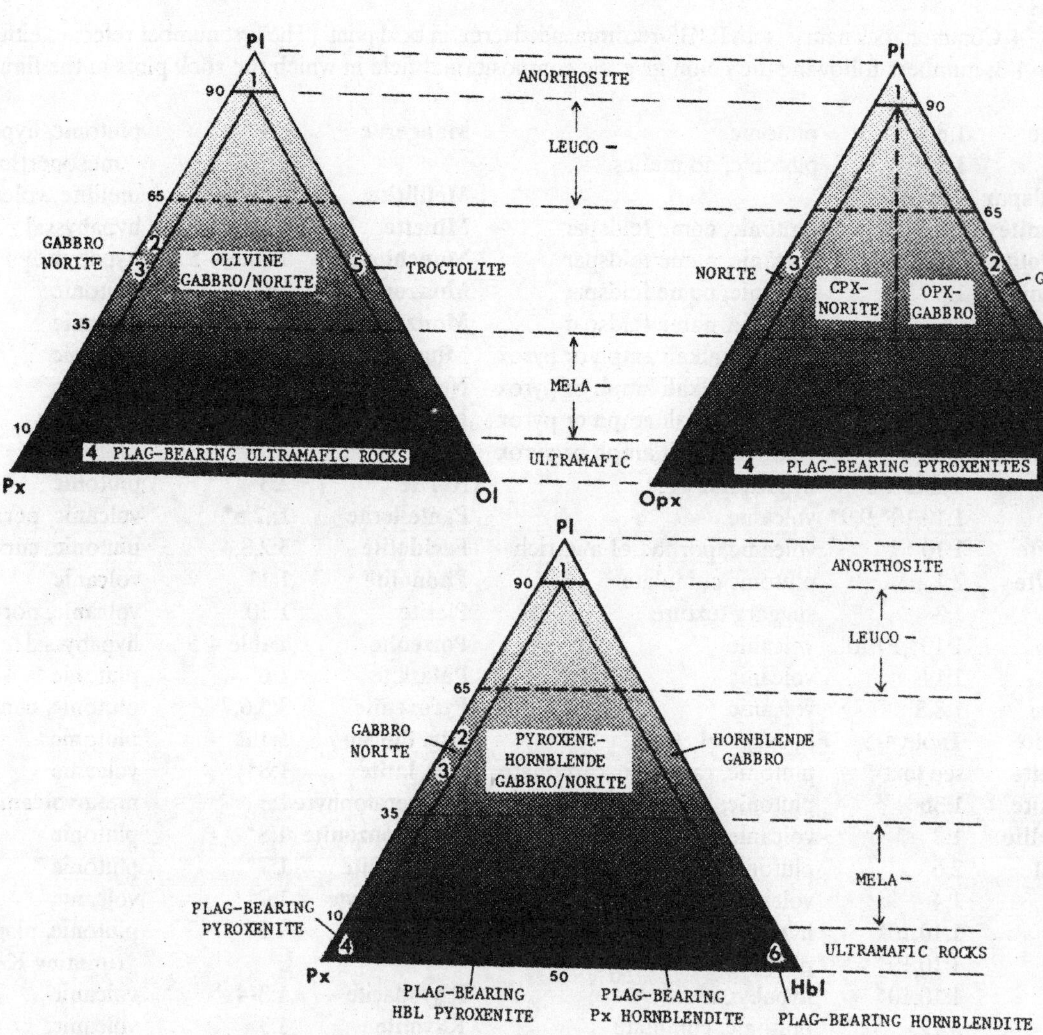

Figures 4-2 IUGS classification of gabbroic rocks. Shading indicates the approximate color index of the rock.

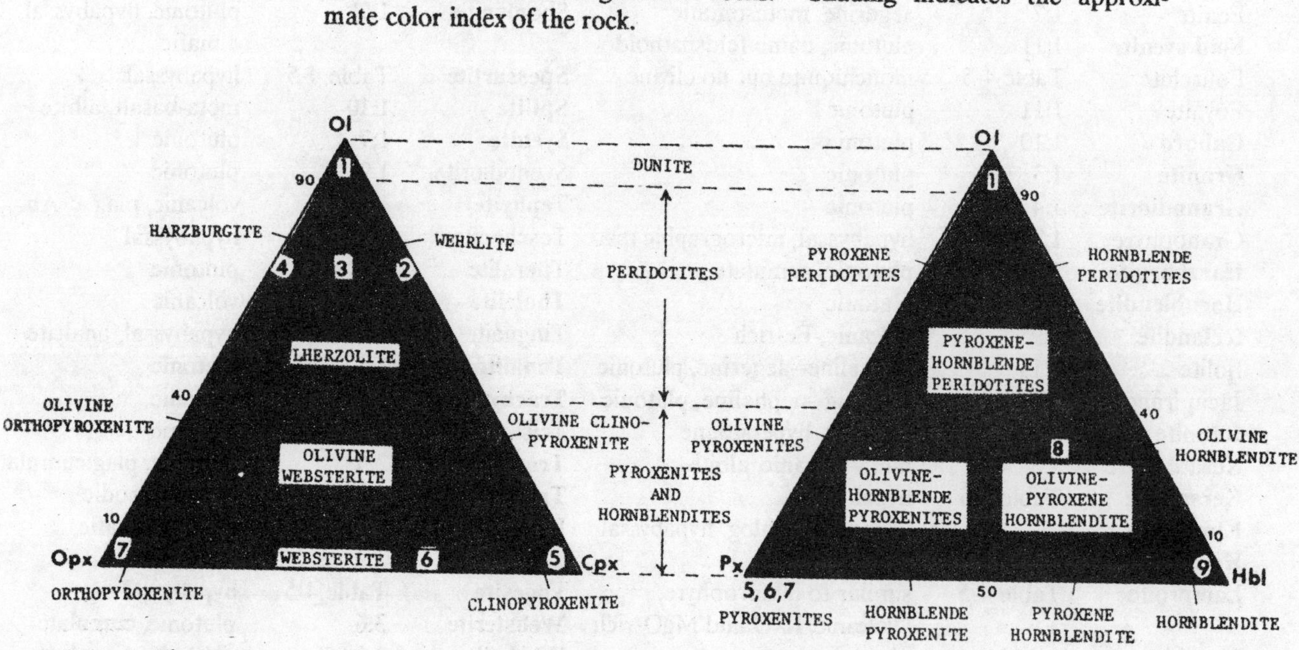

Figure 4-3 Classification of ultramafic rocks.

volcano in northern Tanzania have produced a *natrocarbonatite*, composed essentially of sodium carbonate. This is not known from any other igneous body. Its rarity in the geologic record is undoubtedly due to the high solubility of sodium carbonate in water, which drastically reduces its chances of preservation. Carbonatites in which the carbonate fraction is more than 90% calcite are known as *sovite* if coarse-grained and *alvikite* if finer-grained. Carbonatites in which the carbonate fraction is more than 90% dolomite are known as *beforsite*. Intermediate compositions containing from 10 to 50% dolomite are referred to as *dolomite-calcite carbonatite*, and from 50 to 90% dolomite as *calcite-dolomite carbonatites*. The presence of other minerals, such as olivine, monticellite, diopside, phlogopite, perovskite, pyrochlore, and apatite can be placed in front of the name if their abundance is significant or attention is to be drawn to a particular mineral.

IUGS Classification of Volcanic and Hypabyssal Rocks

Because the modal composition of many volcanic rocks is difficult, if not impossible, to determine, classifications of volcanic rocks have made use of chemical as well as modal data. This has introduced more variables and complicated the classification. As a result, there is no single, generally accepted classification of volcanic rocks as there is of plutonic ones. Nonetheless, the IUGS Subcommission on the Systematics of Igneous Rocks has recommended a classification for adoption (Streckeisen, 1979).

The IUGS recommends that volcanic rocks be classified on the basis of their modes wherever possible, and that the divisions in the classification should correspond to those in the plutonic rock classification. Where modes cannot be obtained, chemical data can be used to classify the rocks, but here again the divisions in the classification should correspond to those for plutonic rocks. The normative calculation provides a simple means of converting chemical data into mineralogical proportions that can be used for classification purposes. Figure 4-1 includes the recommended names for the volcanic rocks in each one of the fields defined by the plutonic rocks. There is a volcanic equivalent for each of the main plutonic rocks, except for those in the ultramafic category and anorthosites. These plutonic rocks are normally formed by the accumulation of crystals from a magma, and they have never existed as a liquid as such. For this reason, they do not appear as lavas. The rare volcanic rock known as *komatiite*, which occurs almost exclusively in Archean terrains, is, however, the volcanic equivalent of peridotite.

Figure 4-1 allows for the simple naming of volcanic rocks based on their modal or normative positions in the diagram. Rocks falling within field 2 should be referred to as alkali-feldspar rhyolites rather than alkali rhyolites, because the latter term has been used to indicate the presence of an alkali pyroxene or alkali amphibole. The same applies to fields 6*, 6, and 6'. The term *rhyodacite* can be used for rocks transitional between fields 3 and 4. Volcanic rocks having compositions similar to tonalite are rare. They can be described as dacite, but the common dacites have the composition of granodiorite. The distinction between basalt and andesite based on plagioclase composition is difficult, because of the problem of determining average plagioclase compositions when crystals are zoned. If an analysis is available, the normative plagioclase composition can be used. Also, the silica content serves to distinguish the two, basalts typically containing less than 52 wt% SiO_2 and andesites more. In general, basalts have color indexes greater than 35, whereas most andesites have lower values. The common andesites of strato volcanoes plot in field 9*, whereas most quartz-bearing basalts fall in fields 10 and 10*. So called alkali basalts, which are associated with many oceanic islands and rift valleys on continents, plot in field 10'. Many volcanic rocks contain glass, which should be mentioned in a rock description. Depending on the percentage of glass present, the rock can be described variously as glass-bearing (0-20% glass), glass-rich (20-50%), glassy (50-80%), or by a special name such as obsidian (for a silica-rich glass), if the glass content is from 80 to 100%.

Most hypabyssal rocks have similar grain sizes to volcanic rocks and are equally difficult to classify. Volcanic rock names are used to describe most of them. Basaltic dikes, if thick enough, may be medium-grained and the rock commonly develops a texture known as *ophitic*, in which laths of plagioclase are embedded in larger crystals of pyroxene or olivine. The synonymous terms *dolerite* and *diabase* are used for such rocks. They plot in fields 10 and 10* of Fig. 4-1.

One distinct group of hypabyssal rocks that have their own nomenclature is the *lamprophyres*, which are melanocratic, porphyritic rocks containing phenocrysts of a hydrous mafic mineral, biotite or amphibole, and possibly clinopyroxene and olivine in a fine-grained groundmass. Feldspars, if present, occur only in the groundmass, but many lamprophyres contain no feldspar at all. These have such low contents of silica that feldspathoids or melilite is present instead. The classification of lamprophyres is given in Table 4-5.

TABLE 4-5 Classification of Lamprophyres after Streckeisen (1979)

Felsic Constituent		Predominant Mafic Mineral			
Feldspar	Foid	biotite diopside augite $\pm$olivine	hornblende diopside augite $\pm$olivine olivine	amphibole - (barkevikite kaersutite) titanaugite $\pm$calcite biotite	melilite biotite $\pm$titanaugite $\pm$olivine
or>pl	-	**Minette**	**Vogesite**		
pl>or	-	**Kersantite**	**Spessartite**		
or>pl	fsp>foid			**Sannaite**	
pl>or	fsp>foid			**Camptonite**	
-	foid			**Monchiquite**	**Polzenite**
	-				**Alnoite**

Because volcanic and hypabyssal rocks form on or near the Earth's surface where they can interact with circulating groundwater, their feldspars are commonly hydrothermally altered to sericite or saussurite (zoisite or epidote) and their mafic minerals to chlorite, serpentine, or talc. Despite this alteration, the rock should be classified according to what the rock was prior to alteration, if this is determinable from textures or relict grains; the prefix *meta-* can be added to indicate that the rock has undergone change. Thus a meta-basalt might consist essentially of albite, epidote, and chlorite. One group of altered rocks for which special terms are used are those formed on the ocean floor. *Spilites* are originally basaltic rock, *keratophyres* originally intermediate volcanic rocks, and *quartz keratophyres* originally silicic volcanic rocks. They have all undergone sodium metasomatism, with the plagioclase being converted entirely into albite, and the mafic minerals are mostly converted to chlorite. Spilites are commonly pillowed, and the tectonic setting of these rocks indicates that they formed on the ocean floor. Their high sodium content may therefore have resulted from interaction with seawater.

More recently (Le Bas, et al., 1986), the IUGS Subcommission on the Systematics of Igneous Rocks has recommended adoption of a simple chemical classification of volcanic rocks based on the two parameters, total alkalis and silica content (TAS). The TAS diagram (Fig. 4-4) is divided into 15 fields by a series of straight lines. To plot a rock in this diagram, the analysis is recalculated to 100% on a H_2O- and CO_2-free basis. Rocks containing more than 2% H_2O^+ (water driven off by heating above 105°C) and more than 0.5% CO_2 are considered altered, and their classification by this scheme may be erroneous. Rocks falling in the trachybasalt field can be further classified as hawaiite if $(Na_2O\text{-}2) > K_2O$ and as potassic trachybasalt if $(Na_2O\text{-}2) < K_2O$. Similarly, the field of basaltic trachyandesite can be divided into mugearite (Na) and shoshonite (K), and the field of trachyandesite into benmoreite (Na) and latite (K). Classification by this scheme is almost totally consistent with that based on the QAPF diagram (Fig. 4-1).

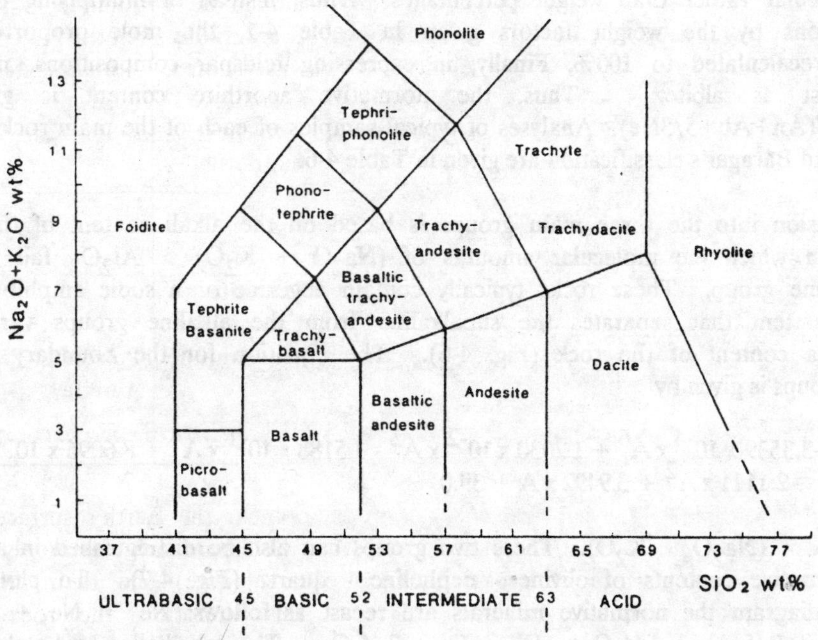

Figure 4-4 Compositional fields of volcanic rocks in terms of total alkalis and silica. (after Le Bas, et al., 1986; published with permission of Oxford University Press)

The Irvine - Baragar Classification of Volcanic Rocks

In recent years the classification of volcanic rocks proposed by Irvine and Baragar (1971) has gained wide acceptance. It sets up divisions between different rock types based solely on common usage; that is, in practice most geologists associate a particular rock name with a certain compositional range. The scheme also incorporates the well-established fact that volcanic rocks fall into a number of distinct genetic series, which can be distinguished by simple chemical parameters. These series have the added significance that they can be correlated with distinct tectonic environments.

Volcanic rocks are classified by Irvine and Baragar into three main groups (Fig. 4-5), the ***subalkaline***, the ***alkaline***, and the ***peralkaline*** (alkali-rich). Most rocks belong to the first two groups, which are each subdivided into two subgroups. Assigning a rock to any one of these groups is based on simple chemical parameters or normative compositions. Before this is done, however, the chemical effects of alteration must be taken into account, if possible. Many volcanic rocks become oxidized, hydrated, or carbonated by hydrothermal activity during burial or during later metamorphism. These chemical changes can seriously affect the normative composition of a rock, which may, in turn, affect its classification. For example, conversion of ferrous iron to ferric during alteration results in smaller amounts of iron silicates being calculated in the norm; this then produces a norm that appears more saturated in silica than was the original rock. This type of alteration, however, can be corrected for, because in many unaltered volcanic rocks there is a strong positive correlation between the TiO_2 and Fe_2O_3 contents. The primary wt% Fe_2O_3 in many volcanic rocks is given approximately by (wt% TiO_2 + 1.5). H_2O and CO_2 are subtracted from the analysis and the total recalculated to 100%. Norm calculations are carried out according to the CIPW rules, but Irvine and Baragar chose to recalculate the normative minerals in terms of molecular rather than weight percentages. Thus, instead of multiplying the mole proportions by the weight factors given in Table 4-1, the mole proportions are simply recalculated to 100%. Finally, in expressing feldspar compositions, nepheline is recast as albite. Thus, the normative anorthite content is given by 100xAn/(An+Ab+5/3Ne). Analyses of typical samples of each of the main rock types in Irvine and Baragar's classification are given in Table 4-6.

Division into the three main groups is based on the alkali content of the rocks. Rocks in which the molecular amounts of $(Na_2O + K_2O) > Al_2O_3$ fall into the peralkaline group. These rocks typically contain aegerine or a sodic amphibole. The alkali content that separates the subalkaline from the alkaline groups varies with the silica content of the rock (Fig. 4-6). The equation for the boundary between these groups is given by

$$SiO_2 = -3.3539 \times 10^{-4} \times A^6 + 1.2030 \times 10^{-2} \times A^5 - 1.5188 \times 10^{-1} \times A^4 + 8.6096 \times 10^{-1} \times A^3 - 2.1111 \times A^2 + 3.9492 \times A + 39.0$$

where $A = (Na_2O + K_2O)$. These two groups can also be distinguished in a plot of the normative contents of olivine - nepheline - quartz (Fig. 4-7). To plot a rock in this diagram the normative minerals are recast as follows: Ne' = Ne + 3/5 Ab, Q' = Q + 2/5 Ab + 1/4 Opx, Ol' = Ol + 3/4 Opx. The subalkaline rocks plot on the quartz side of the boundary line, whereas the alkaline ones plot on the nepheline side of it.

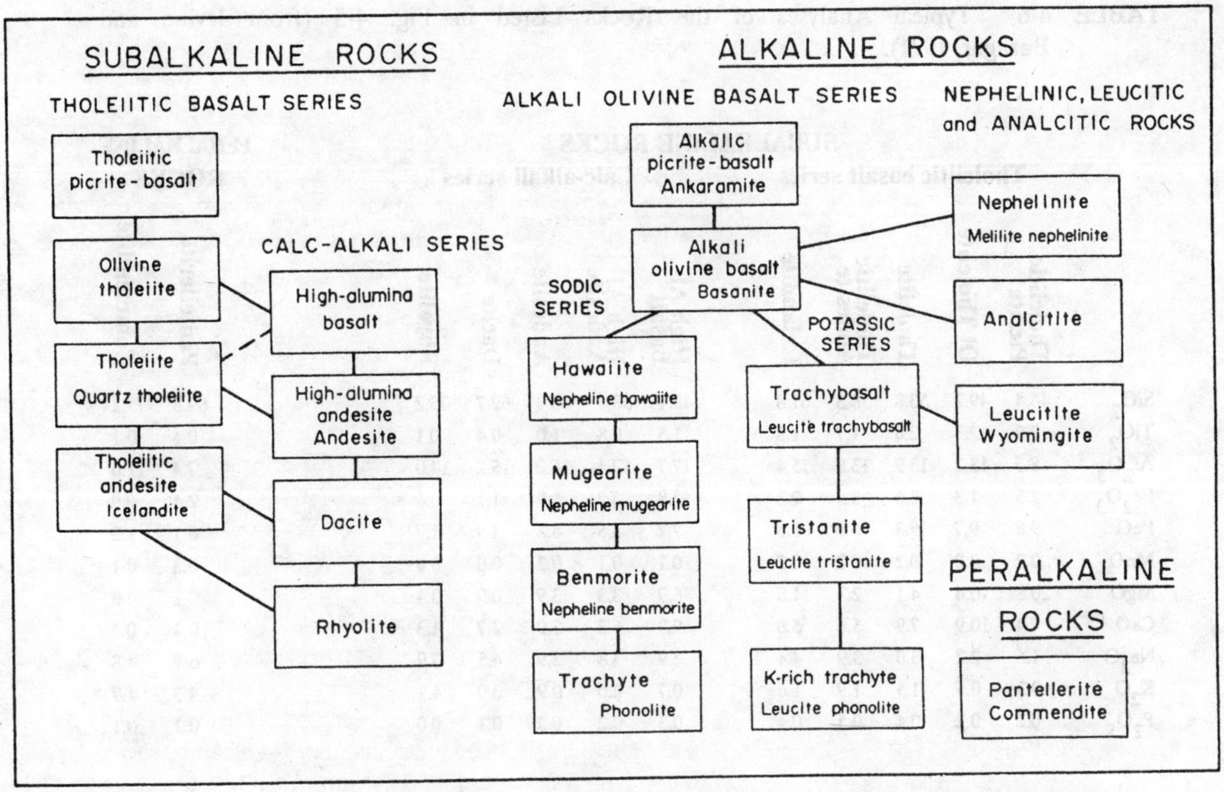

Figure 4-5 General classification scheme for the common volcanic rocks (after Irvine and Baragar, 1971). Lines joining boxes link commonly associated rocks. The small print within the boxes refers to variants of the main rock. (Published with permission of Canadian J. Earth Sci.)

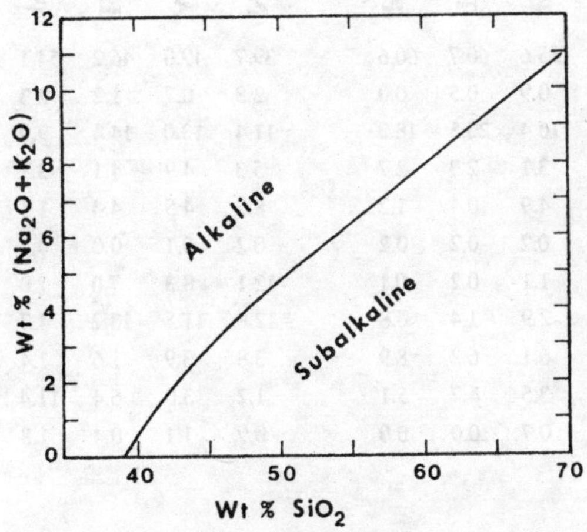

Figure 4-6 Alkalis-silica plot with line separating fields of alkaline and subalkaline rocks. (after Irvine and Baragar, 1971).

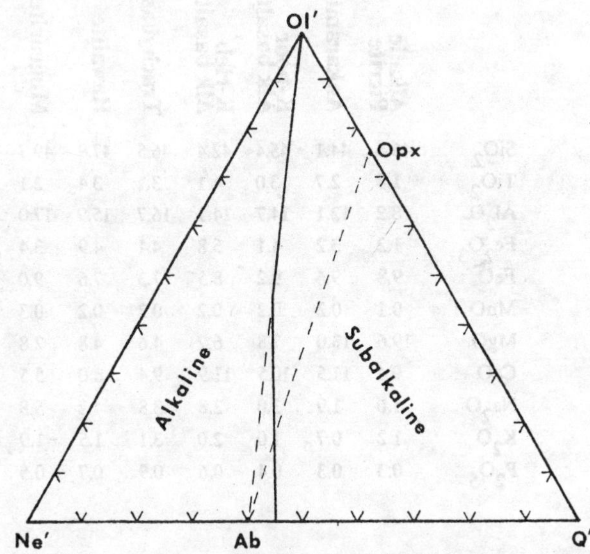

Figure 4-7 Ol'-Ne'-Q' projection with line separating fields of alkaline and subalkaline rocks. See text for explanation. Plot in % cation equivalents. (after Irvine and Baragar, 1971)

TABLE 4-6 Typical Analyses of the Rocks Listed in Fig. 4-5 (from Irvine and Baragar, 1971).

	SUBALKALINE ROCKS: Tholeiitic basalt series					SUBALKALINE ROCKS: Calc-alkali series					PERALKALINE ROCKS	
	Tholeiitic Picrite	**Ol-Tholeiite**	**Tholeiite**	**Tholeiitic Andesite**	**Icelandite**	**High Al Basalt**	**High Al Andesite**	**Andesite**	**Dacite**	**Rhyolite**	**Pantellerite**	**Commendite**
SiO_2	46.4	49.2	53.8	58.3	61.8	49.1	58.6	60.0	69.7	73.2	69.8	75.2
TiO_2	2.0	2.3	2.0	1.7	1.3	1.5	0.8	1.0	0.4	0.1	0.4	0.1
Al_2O_3	8.5	13.3	13.9	13.8	15.4	17.7	17.4	16.0	15.2	14.0	7.4	12.0
Fe_2O_3	2.5	1.3	2.6	3.4	2.3	2.8	3.2	1.9	1.1	0.6	2.4	0.9
FeO	9.8	9.7	9.3	6.5	5.8	7.2	3.5	6.2	1.9	1.7	6.1	1.2
MnO	0.2	0.2	0.2	0.2	0.2	0.1	0.1	0.2	0.0	0.0	0.3	0.1
MgO	20.8	10.4	4.1	2.3	1.8	6.9	3.3	3.9	0.9	0.4	0.1	0.0
CaO	7.4	10.9	7.9	5.6	5.0	9.9	6.3	5.9	2.7	1.3	0.4	0.3
Na_2O	1.6	2.2	3.0	3.9	4.4	2.9	3.8	3.9	4.5	3.9	6.7	4.8
K_2O	0.3	0.5	1.5	1.9	1.6	0.7	2.0	0.9	3.0	4.1	4.3	4.7
P_2O_5	0.2	0.2	0.4	0.5	0.4	0.3	0.2	0.2	0.1	0.0	0.2	0.1

	ALKALINE ROCKS: Alkali olivine basalt series											ALKALINE ROCKS: Nephelinites etc.			
	Alkalic Picrite	**Ankaramite**	**K-poor Alk basalt**	**K-rich Alk basalt**	**Trachybasalt**	**Hawaiite**	**Mugearite**	**Tristanite**	**Benmorite**	**Trachyte**	**Phonolite**	**Nephelinite**	**Analcitite**	**Leucitite**	**Wyomingite**
SiO_2	46.6	44.1	45.4	42.4	46.5	47.9	49.7	55.8	55.6	60.7	60.6	39.7	49.0	46.2	54.1
TiO_2	1.8	2.7	3.0	4.1	3.1	3.4	2.1	1.8	0.9	0.5	0.0	2.8	0.7	1.2	2.3
Al_2O_3	8.2	12.1	14.7	14.1	16.7	15.9	17.0	19.0	16.4	20.5	18.3	11.4	13.0	14.4	9.9
Fe_2O_3	1.2	3.2	4.1	5.8	4.1	4.9	3.4	2.6	3.1	2.3	2.7	5.3	4.9	4.1	3.1
FeO	9.8	9.6	9.2	8.5	7.3	7.6	9.0	3.1	4.9	0.4	1.2	8.2	4.5	4.4	1.5
MnO	0.1	0.2	0.2	0.2	0.2	0.2	0.3	0.1	0.2	0.2	0.2	0.2	0.1	0.0	0.1
MgO	19.6	13.0	7.8	6.7	4.6	4.8	2.8	2.0	1.1	0.2	0.1	12.1	8.3	7.0	7.0
CaO	9.4	11.5	10.5	11.9	9.4	8.0	5.5	4.5	2.9	1.4	0.8	12.8	11.5	13.2	4.7
Na_2O	1.6	1.9	3.0	2.8	3.8	4.2	5.8	5.2	6.1	6.2	8.9	3.8	3.9	1.6	1.4
K_2O	1.2	0.7	1.0	2.0	3.1	1.5	1.9	4.1	3.5	6.7	5.1	1.2	3.0	6.4	11.4
P_2O_5	0.3	0.3	0.4	0.6	0.9	0.7	0.5	0.4	0.7	0.0	0.0	0.9	1.1	0.4	1.8

The subalkaline rocks are divided into the *calc-alkali* and *tholeiitic* series on the basis of their iron contents in the AFM plot (Fig. 4-8), where A = Na_2O + K_2O, F = FeO + $0.8998xFe_2O_3$, and M = MgO (all in wt%). This plot distinguishes intermediate members of these series very well, but at the mafic and felsic ends there is considerable overlap. Calc-alkali basalts and andesites, however, contain 16 to 20% Al_2O_3, which is considerably more than occurs in tholeiitic basalts and andesites which contain from 12 to 16%. At the extreme felsic end there is no satisfactory way of distinguishing calc-alkali and tholeiitic members; thus all granitic rocks are assigned to the calc-alkali series.

The alkaline rocks are divided into the *alkali olivine basalt* series and the *nephelinitic - leucitic - analcitic* series. Rocks of the latter series typically contain less than 45% SiO_2, have normative color indices greater than 50, and may contain normative leucite.

The naming of rocks within the various subgroups is based on normative plagioclase composition and on normative color index. In the various subalkaline series, the rocks range from *basalt* through *andesite* and *dacite* to *rhyolite* with decreasing normative anorthite content and decreasing normative color index (Fig. 4-9). Two series of rock names are used for the alkaline rocks, depending on whether they are sodic or potassic. This division is made on the basis of the normative feldspar composition (Fig. 4-10). With decreasing normative anorthite content, the sodic series passes from *alkali basalt* through *hawaiite, mugearite*, and *benmoreite* to *trachyte* (Fig. 4-11a), whereas the potassic series passes from *alkali basalt* through *trachybasalt* and *tristanite* to *trachyte* (Fig. 4-11b). At the mafic end of all of these series, basalts containing more than 25% normative olivine are named *picrites*; these rocks contain abundant phenocrystic olivine. *Ankaramites*, which belong to the alkaline group, contain abundant augite phenocrysts which cause the norm to have more than 20% clinopyroxene. Basalts containing more than 5% normative nepheline are named either *basanite* if they contain modal nepheline, or *basanitoid* if nepheline is not visible. Finally, nepheline-bearing trachyte is known as *phonolite*.

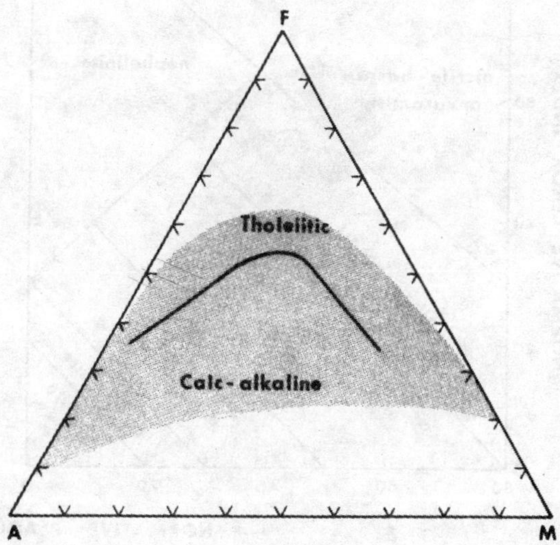

Figure 4-8 AFM plot showing line separating fields of tholeiitic and calc-alkaline rocks as proposed by Irvine and Baragar (1971). A = Na_2O+K_2O; F = $FeO+0.8998Fe_2O_3$; M = MgO, all in weight percent.

Figure 4-9 Irvine and Baragar's (1971) subdivision of the subalkaline rocks in a plot of normative color index versus normative plagioclase composition. Plot in % cation equivalents. Normative color index = ol + opx + cpx + mt + il + hm. Normative plagioclase composition = 100An/(An + Ab + 5/3Ne); this converts nepheline into albite.

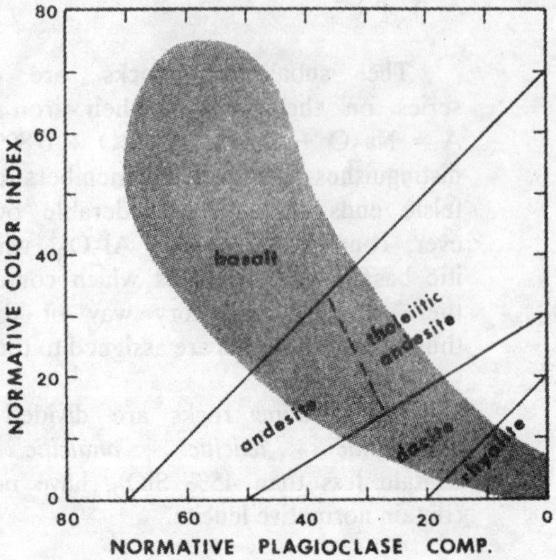

Figure 4-10 Plot of normative An - Ab' - Or with dividing line separating sodic and potassic alkaline rocks. Ab'= Ab + 5/3Ne. Plot in % cation equivalents. (after Irvine and Baragar, 1971).

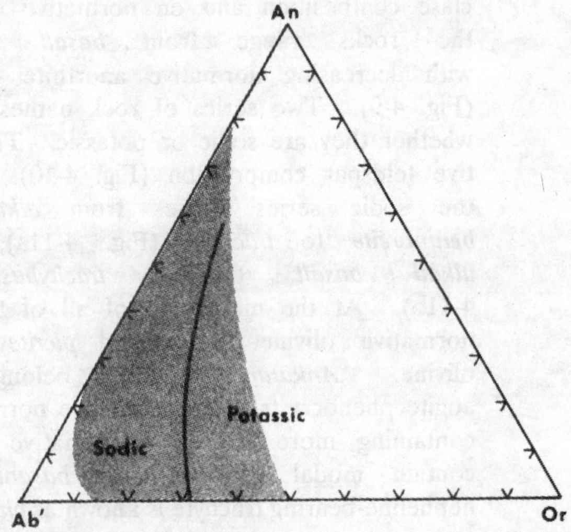

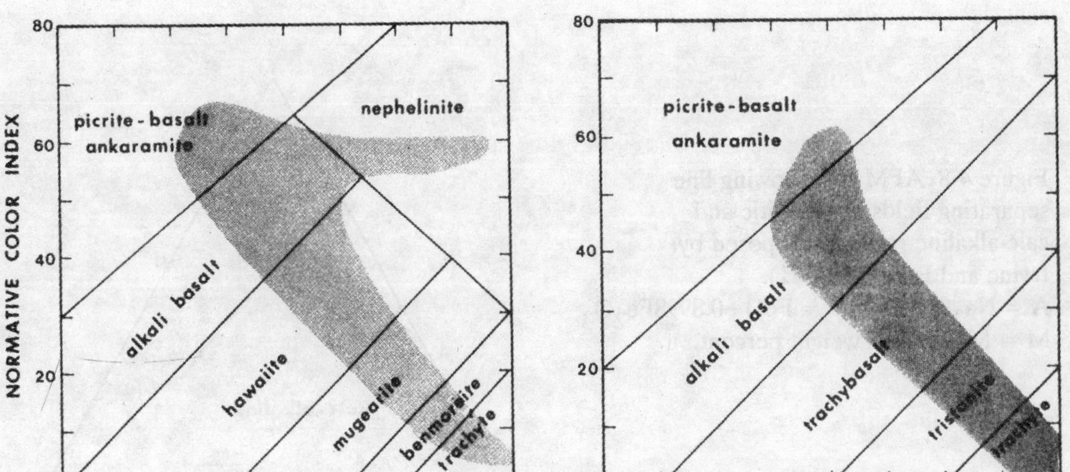

Figure 4-11 Plots of normative color index versus normative plagioclase composition (defined as in Fig. 4-9) for (a) sodic alkaline rocks and (b) potassic alkaline rocks. (after Irvine and Baragar, 1971)

Although this classification is descriptive, it has considerable genetic significance. The calc-alkali series is characteristic of orogenic belts and gives rise to the volcanic rocks of island arcs. These rocks are clearly related to subduction zones. The tholeiitic rocks are prominently developed in zones of crustal extension where they commonly develop thick sequences of flood basalts. They constitute the major rock type along oceanic ridges (MORB = mid-ocean ridge basalt) and on many large oceanic islands, such as Hawaii. Alkali olivine basalts and associated rocks commonly occur in areas of continental rifting and in regions overlying deeply subducted plates; they also occur at intra-plate hot spots in both oceanic and continental regions.

Igneous Rock Names

Despite the relatively small number of rock names proposed in the IUGS and Irvine and Baragar classifications, the petrologic literature contains hundreds of rock names. Although most of these are no longer used, it is necessary at least to be able to find their definitions in order to read the literature. The list in Table 4-4 includes the most commonly encountered rock names. They have been defined, if possible, by relating them to the IUGS classification with a pair of numbers. The first number, 1, 2, and 3, refers to Figs. 4-1, 4-2, and 4-3, and the second number indicates the field in which the rock plots in the particular figure. Thus a jotunite (1:9) is a hypersthene monzodiorite and plots in Fig. 4-1, field 9. Some rock names are based on textural features, in which case these are briefly stated. The object of this tabulation is not to present a catalogue of names to be memorized but simply to provide a convenient list to which reference can be made quickly. The important rock names, including for example those in the IUGS classification, have been placed in bold print, and their definitions should be learned.

REFERENCES

Irvine, T. N., and Baragar, W. R. A., 1971, A guide to the chemical classification of common volcanic rocks: Canadian J. Earth Sci., v.8, p. 523-548.

Johannsen, A., 1931, *A Descriptive Petrography of the Igneous Rocks. Vol. I, Introduction, Textures, Classification and Glossary*, University of Chicago Press, Chicago, 267 p.

Le Bas, M. J., Le Maitre, R. W., Streckeisen, A., and Zanettin, B., 1986, A chemical classification of volcanic rocks based on the total alkali-silica diagram: J. Petrol., v. 27, p. 745-750.

Streckeisen, A., 1976, To each plutonic rock its proper name: Earth Sci. Rev., v. 12, p. 1-33.

Streckeisen, A., 1979, Classification and nomenclature of volcanic rocks, lamprophyres, carbonatites, melilitic rocks: Recommendations and suggestions of the IUGS Subcommission on the Systematics of Igneous Rocks: Geology, v.7, p.331-335.

5 Textures and Structures of Igneous Rocks

The terms *texture* and *structure* are commonly used interchangeably, but there is a distinction between them. *Texture* refers to the way in which individual grains relate to grains immediately surrounding them (for example, IT 20). Texture deals with small-scale features seen in hand specimens or under the microscope, such as the degree of crystallinity, grain size, grain shape, and crystal intergrowths. *Structure* refers to the way in which aggregates of grains or small bodies of one rock relate to another (for example, IT 41). Structure deals with larger features that are seen at the scale of an outcrop, and includes, for example, flow structures, layering, mineralogical zonation, and xenoliths. Textures are useful indicators of cooling and crystallization rates and of phase relations between minerals and magma at the time of crystallization. Structures, on the other hand, can provide information on processes active during the formation of rocks and on mechanisms of differentiation.

TEXTURES

1. Degree of Crystallinity

Magma, if cooled rapidly enough, quenches to glass. This is particularly true of magmas with high silica contents, which quench to form the common *obsidian*. The identification of glass in a rock can therefore serve as an indication not only of cooling rate, but possibly of composition. The crystallinity of a rock can vary from *holohyaline* (totally glassy), through *hypohyaline* (IT 10) or *hypocrystalline* (glass and crystals), to *holocrystalline* (totally crystalline)(IT 13).

2. Grain Size

The grain size of an igneous rock is determined largely by the rate of cooling of the magma, but in some plutonic bodies the volatile content of the magma may play a more important role. Volatiles increase diffusion rates, which decrease the number of nuclei formed and increase the grain size. Grain sizes are divided into *fine*, *medium*, and *coarse* (IT 5-8) on the basis of whether a microscope, hand lens, or unaided eye is adequate to identify the constituent minerals. Conventionally the divisions between these have been placed at 1 mm and 5 mm respectively. If a rock is so fine-grained that individual crystals cannot be discerned without a microscope, it is said to be *aphanitic*. If the grains of the essential minerals are visible to the unaided eye, the rock is *phaneritic*. If the rock is exceptionally coarse-grained, which typically results from it having crystallized from a volatile-rich magma, it is described as *pegmatitic*.

Grains in a rock may not all be of the same size. Indeed, most volcanic and near surface igneous rocks have grains of several different sizes. A rock containing grains of distinctly different sizes is said to be *porphyritic* (IT 9). The larger crystals, which are referred to as *phenocrysts*, are surrounded by finer-grained material known as *groundmass*. If the groundmass is glassy, the rock is said to have a *vitrophyric* texture (IT 10). A *mesostasis* is a fine aggregate of minerals or glass that fills the interstices between the grains of the groundmass. It occurs only in volcanic or near-surface intrusive rocks. Rocks containing no phenocrysts are rare, and are referred to as *aphyric*. Phenocrysts may cluster and grow together to form a *glomeroporphyritic* texture.

3. **Grain Shapes**

Grains can be bounded entirely by crystal faces, in which case they are said to be *euhedral* (IT 1); if they have a few crystal faces, they are *subhedral* (IT 3), and if they have none, they are *anhedral* (IT 2). Whether a grain develops crystal faces by growing freely in a melt or is forced to take on the shape imposed by surrounding crystals depends to a large extent on the stage at which the mineral crystallizes. The determination of grain shape, then, can provide information on the sequence of crystallization. If most of the grains in a rock are euhedral, the rock is said to be *panidiomorphic*; when most are anhedral, it is *allotriomorphic*. Most rocks, however, have some grains with crystal faces and are described as *hypidiomorphic* (IT 3). Equidimensional grains are said to be *equant*, whereas flattened or plate-like ones are said to be *tabular*; if grains are elongated and bounded by prism faces, they are *prismatic* (IM 35), if blade-like, they are *laths* (IT 16-18), and if needle-shaped, they are *acicular* (IT 11). A *granitic* texture indicates a rock is hypidiomorphic and equigranular. An *aplitic* texture indicates that a rock is fine-grained, allotriomorphic, and equigranular. Because of an aplite's resemblance to fine sugar, it is also referred to as *saccharoidal*.

4. **Textures of Glassy or Fine-Grained Rocks**

Rapidly cooled magma can form glass, which, being supercooled liquid, is metastable and in time may crystallize, especially in the presence of water. The crystals that grow in glass are quite unlike those formed by normal growth from a liquid. Individually discernible crystals are not formed, and even under the microscope a birefringent fibrous meshwork is all that is normally visible. This type of crystallization is known as *devitrification*. Fibres commonly grow perpendicular to cracks in the glass or radiate outward from phenocrysts to produce spherical bodies known as *spherulites* (IT 10). Many devitrified glasses also have a striking network of curved and concentric fractures referred to as *perlitic* cracks.

Small crystals that have grown in rapidly cooled magma, but prior to complete solidification, are termed *microlites* (IT 10). Because their growth is rapid and diffusion in silicate melts is slow, they grow with *dendritic* habit (IM 9), and are characterized by *skeletal* forms with considerable open space. Pyroxene microlites are commonly acicular, and those of plagioclase are narrow laths with ends resembling swallow tails or tuning forks (IM 55). Although most microlites are visible only under the microscope, those of olivine in Archean ultramafic lavas form sheaves of millimeter-wide crystals that have lengths of tens of centimeters (IM 10). The sheaves, which nucleate at the rapidly cooled tops of flows, radiate downward, and produce a texture resembling inverted tufts of grass; this texture is consequently given the name *spinifex* (a type of grass).

Many mafic dike rocks of the lamprophyre family contain spherical bodies of felsic material known as *ocelli* (IT 53). Unlike spherulites, these consist of relatively coarse crystals of minerals that typically form the mesostasis of the surrounding rock. The *ocellar* texture has been interpreted as resulting from late-stage liquid flowing into gas cavities or magma splitting into immiscible felsic and mafic liquids. The term *variolite* has been used as synonymous with ocelli, but it has also been used to describe spherulites.

Unquestionable examples of *immiscible liquids* are found in the mesostasis of many tholeiitic basalts, where small micron-size droplets of brown, iron-rich silica glass occur in clear silica-rich glass (IT 50). The iron-rich droplets may crystallize to spherical aggregates of pyroxene, magnetite, and apatite if not cooled rapidly enough, but the silica-rich host typically solidifies to glass (IT 51). Immiscible sulfide liquid droplets also commonly occur in the glassy mesostasis of basalts. In thin sections these form opaque spheres (IT 55-56).

5. Flow Textures

If flow continues during the cooling and crystallization of lava, alignment of crystals may result. This is particularly true of platy feldspar laths, which produce a *trachytic* texture (IT 16-17), named from its common occurrence in the rock type trachyte. A similar texture involving the more blocky feldspar laths in plutonic rocks is referred to as *trachytoidal* (IT 18).

6. Intergrowth Textures

The minerals in an igneous rock may have crystallized simultaneously or sequentially, or a mineral may have formed at the expense of another. A number of important textures provide evidence for each of these cases.

A *poikilitic* texture refers to small granular crystals randomly enclosed within larger crystals of another mineral (IT 33). Unlike the porphyritic texture, the large crystals, which are known as *oikocrysts*, have no crystal faces and appear to have crystallized late. Furthermore, the granular shape of the enclosed minerals suggests that resorption or reaction with the surrounding mineral may have been involved in the development of this texture.

Closely related to the poikilitic texture is the *ophitic* one where randomly oriented laths of plagioclase are enclosed in larger crystals of pyroxene or olivine (IT 27). In contrast to the poikilitic texture, the enclosed laths are euhedral, with no sign of reaction or resorption. Indeed, this texture most likely results from the coprecipitation of plagioclase with pyroxene or olivine. The term *subophitic* is used where the olivine or pyroxene crystals are smaller, and thus cannot enclose as many plagioclase laths (IT 28). A precise difference between these terms can be made on the basis of the ratio of the numbers of plagioclase crystals to olivine or pyroxene crystals. If this ratio is greater than one, the texture is ophitic, and if equal to one, it is subophitic. If the ratio falls below one, the rock has an *intergranular* texture (IT 29). With the subophitic texture, a single olivine or pyroxene crystal will, on average, fill each triangular space between plagioclase laths; with the intergranular texture, several olivine or pyroxene grains will fill this space. The change from intergranular through subophitic to ophitic results from slower cooling and slower nucleation rates. This textural sequence is commonly found in progressing from the margins towards the center of wide diabase dikes. If the cooling rate is very fast, the interstices between the plagioclase laths may be quenched to a glass to form an *intersertal* texture.

Early-crystallizing minerals can become segregated and concentrated to form rocks that have distinctly different compositions from the magma from which they formed. This accumulation of minerals produces a *cumulus* texture (IT 43-44), and the rock is referred to as a *cumulate*, with a prefix added for the mineral or minerals that accumulated. Cumulus minerals tend to be euhedral, and rest against one another in a loosely packed fashion. Intercumulus liquid trapped between cumulus grains crystallizes to its constituent minerals, which may bear a poikilitic or ophitic relation to the cumulus minerals. Commonly the composition of the intercumulus liquid changes prior to solidification as a result of either diffusion or actual flushing with liquid of a different composition. Replenishment of the intercumulus liquid in those components that have already been depleted in forming the cumulus grains can lead to additional growth on the cumulus crystals to produce an *adcumulus* texture (IT 45). When carried to an extreme, a monomineralic rock can result.

Quartz and alkali feldspar in granites, particularly pegmatitic ones, commonly exhibit an intergrowth texture consisting of triangular-, wedge-, and hexagonal--shaped quartz crystals distributed throughout large single crystals of alkali feldspar. The shapes and distribution of the quartz crystals resemble cuneiform writing, and thus the texture is referred to as *graphic* (IT 20-21). In more rapidly cooled granitic magma a similar texture is developed, but it is less regular and is finer-grained (visible normally only under the microscope). This is referred to as a *granophyric* or *micrographic* texture (IT 22-23). A still less regular, worm-like intergrowth of quartz in oligoclase that extends into potassium feldspar crystals is known as *myrmekite* (IT 24-25). Although some myrmekite may be of igneous origin and related to the granophyric texture, most probably forms through a solid state replacement reaction.

At high temperatures, alkali feldspars form a complete solid solution series, but on slow cooling they separate into sodium- and potassium-rich feldspars. This exsolution develops the *perthitic texture* (IT 47), in which lamellae of albite occur in a potassium-feldspar host crystal. The term *antiperthite* has been used where the lamellae are of potassium feldspar and the host is plagioclase.

Granites formed at high temperature contain a single alkali feldspar which may exsolve on cooling to form perthite. Some granites, however, in particular those with high water contents, crystallize below the temperature at which alkali feldspars form a continuous solid solution series. In these, the alkali feldspars must crystallize from the magma as separate sodium- and potassium-rich phases; that is, two distinct feldspars are present, in contrast to the single homogeneous alkali feldspar formed initially in the high-temperature granite. Because the line in a phase diagram marking the temperature below which exsolution takes place is called a *solvus* (IT 46), granites containing a single high-temperature feldspar (later possibly exsolving to form perthite) are known as *hypersolvus granites* (IT 47), and those lower-temperature ones containing separate crystals of potassium feldspar and albite are known as *subsolvus granites* (IT 48). Note that the two feldspars in the subsolvus granite may both become perthitic on cooling, but the primary two-feldspar nature of this rock will remain evident.

7. Reaction Texture

Reaction textures are special intergrowths where the distribution of grains is clearly the result of reaction. Reactions may occur between crystals and magma to produce *coronas* (IT 34-35), or between crystals alone to produce *kelyphitic* rims (IT 39). The distinction between these is not always simple or possible to make, but generally coronas form coarse-grained rims around early-formed minerals; they commonly produce a poikilitic texture. Kelyphitic rims, on the other hand, tend to be fine-grained, and may consist of several concentric shells with the reaction products forming worm-like or fibrous intergrowths. Clear examples of both are known.

Olivine crystals, for example, react with cooling tholeiitic magma to produce coronas of orthopyroxene. In slowly cooled magma this orthopyroxene may form large oikocrysts enclosing rounded olivine grains (IT 33). Under high pressures, olivine and calcic plagioclase react in the solid state to form kelyphitic rims of hypersthene, pargasitic amphibole, and spinel (IT 39).

Minerals belonging to continuous reaction series exhibit evidence of reaction by forming zoned crystals. The plagioclase in gabbros, for example, is *normally zoned* from high-temperature calcic cores to low temperature more sodic rims. If this sequence is reversed, the crystals are said to be *reversely zoned*. In some rocks, especially diorites and andesites, the composition of the plagioclase may fluctuate back and forth to form *oscillatory zoning* (IM 54). The normal sequence of crystallization in granites can result in early-formed plagioclase being rimmed by alkali feldspar. In granites with a *rapakivi* texture, however, early-crystallizing potassium feldspar is mantled by oligoclase.

STRUCTURES

1. Structures in Volcanic Rocks

Structures developed on the surface of lava flows may preserve a record of the manner in which a lava moves. Lavas can travel at velocities ranging from meters per second for basalt to meters per month for rhyolite. The surface of any lava, regardless of its composition, cools rapidly by radiation, but heat from the flow's interior must be transferred to the surface by conduction, an extremely slow process. The interior of a lava will consequently remain molten and possibly continue flowing long after the surface has solidified. This differential movement tends to disrupt the crust, and, depending on flow velocities, produces a number of structures.

Basalt, which has a viscosity similar to that of tar, commonly flows rapidly enough that it is able to deform its surface during cooling. The viscous surface layer becomes wrinkled into arcuate ridges that are convex in the direction of flow. Because these ridges resemble rope, this surface structure is referred to as *ropy lava*; the Hawaiian term *pahoehoe* is also used. Molten basalt may issue forth onto the surface of a flow from beneath a solid crust. As it does so, irregularities on the underside of the solid crust impress grooves in the newly emerging surface to form lines that parallel the direction of flow. Where this lineated surface becomes wrinkled into a ropy structure, the lines are approximately transverse to the ridges.

Once basaltic lava begins to move more slowly, its surface cools and becomes rigid. Continued movement of the still-molten interior breaks up this cooler surface layer into large slabs and eventually into small clinker-like fragments, the surfaces of which are extremely rough with many broken glassy edges. With continued movement, a large fraction of a lava flow can be converted into this rubble, which continues to move, albeit slowly, by material avalanching down the steep front face of the flow to be overridden by more rubble in a conveyor-like motion. Such clinker-like lava is referred to by its Hawaiian name, *aa*.

The surface of andesite flows may also be covered with rubble, but these fragments are typically larger than those in aa flows (head-size in contrast to fist-size in aa) and have smoother surfaces. These are referred to as *blocky* lava flows. The surface of many rhyolite flows also consists entirely of broken fragments of glass up to a meter or more across.

It is not uncommon in flows of basalt and basaltic andesite for the molten interior to drain out from beneath the solid crust, leaving behind a *lava tube*. The dimensions of these depend on the thickness and length of the flow and are commonly many meters in diameter and hundreds of meters in length. The crust above a tube may sag, forming an elongate depression, or even collapse, which then provides access to the tube (most are plugged with lava at their lower end). Stalactite-like dribbles of lava hang from the ceiling, and horizontal ridges on the walls mark periods during which the depth of the lava remained relatively constant during the draining. Tubes may divide and rejoin around pillars, which provide support for the roof. The upper surface of a lava flow above such a pillar may be domed upward to form a *schollendome*. If sagging of the crust into the surrounding tube is sufficiently great, the crest of the schollendome may be marked by wide tensional fractures. Domes and *pressure ridges* can also form on the surface of flows independent of lava tubes as a result of increased flow beneath a solid crust which causes cracking and upwelling of lava onto the surface.

The surface of basaltic lava extruded in water is quenched almost instantly to a glass, which may temporarily prevent the lava from advancing. Eventually, however, this selvage may crack and a tongue of hot, molten lava can momentarily advance into the water where it is quenched to a sack-like body known as a *pillow* (IT 14). By successive cracking and budding the basalt advances as a *pillow lava*. The presence of pillows is clear evidence of extrusion beneath water. Pillows range from small bodies the size of a soccer ball to ones the size of a mattress. The larger the pillow, the flatter it tends to be, because its interior remains molten longer allowing more time for sagging and flattening. Where pillows are piled one on top of another the base of pillows exhibits cusp-like tails that sag into the space between underlying pillows; in contrast, the upper surface of a pillow is normally smooth and convex upward. This asymmetry in pillow shape provides a useful means of determining tops in pillow lavas that have been turned up on end during tectonism.

In addition to surface structures, many lavas, especially the more siliceous ones, have internal structures indicative of flow. Many glassy or partially glassy lavas exhibit prominent *flow layering*, which consists of millimeter- to centimeter-thick layers of varying color and texture. Layers rich in glass tend to be dark and have a vitreous luster, whereas those containing microlites, small gas bubbles, or spherulites are lighter colored and have a resinous luster; the microlites commonly exhibit a trachytic texture. The layering can also result from variations in the abundance of phenocrysts. Much of this layering is present in the lava before it leaves the volcanic vent, and as the lava spreads laterally, the layering can be contorted into *flow folds*, the attitude of which preserve a record of the lava's direction of movement.

Dissolved gases in magma are likely to come out of solution following eruption, because of their decreased solubility at low pressure. In addition, crystallization of anhydrous minerals augments this effect by increasing the concentration of volatiles in the residual liquid.

Bubbles of exsolved gas in lava are called *vesicles* (IT 12), and rocks containing these are said to be *vesicular.* The diameter of vesicles is typically measured in millimeters. Gas cavities larger than a centimeter are likely formed by the coalescence of vesicles or by gas that has risen into the lava from beneath the flow. These are termed *vugs* or *geodes* if they have a subspherical shape and are lined with banded chalcedony and inward projecting quartz crystals. Vesicles are initially spherical, but they may be deformed into ellipses by the flow of lava. Vesicles may collapse into irregular convoluted shapes once exsolution of gas is essentially complete and the gas in the vesicles begins to cool and contract. Vesicles in aa flows typically have irregular shapes, whereas those in pahoehoe flows are spherical.

Vesicles may be so abundant as to produce a rock resembling a froth. This is particularly common at the top of flows where vesicles collect through buoyant rise. The dark and commonly oxidized rock formed in this way on the surface of basaltic flows is referred to as *scoria*, whereas the light colored equivalent on rhyolitic flows is termed *pumice*.

Exsolution of gas from lava that initially contains very little gas may occur at such a late stage in the solidification process that typical vesicles cannot form. Instead, the gas may occupy cavities between crystals; that is, the gas cavity will be bounded by planar crystal faces. Such late stage exsolution of gas occurs in some basalts forming what is referred to as a *dictytaxitic* texture (IT 13).

Vesicles at the base of many basaltic flows extend into the lava as long tubes. If these are later filled with light-colored secondary minerals they resemble the stems of old-fashioned clay pipes, and consequently are referred to as *pipe stem vesicles* or simply *pipe vesicles*. Although commonly interpreted as having formed by gas rising into a flow, they are more likely the result of nucleation and growth of gas bubbles on the solidification front as it advances into the lava. In pillows, for example, pipe vesicles are found extending not only up from the base, but down from the top and in from the sides, where buoyancy could not possibly be a factor in their growth (IT 14). Many pipe vesicles extending up from the base of lavas are tilted at their upper end in the direction of flow of the lava. These provide a useful means of determining flow directions.

Basaltic lavas, especially those that are vesicular at their base, may have cylinders of vesicular basalt, several centimeters in diameter and typically several meters in length, extend up into the flow. These are known as *vesicle cylinders*; they are produced by the buoyant diapiric rise of less dense, vesicular lava into more dense, massive lava. They are commonly capped by a large vug lined with vesicular basalt. These vugs tend to be hemispherical with a flat, horizontal base. Vesicle cylinders, like pipe vesicles, may be tilted in the direction of flow of the lava.

Vesicles may become filled with low-temperature secondary minerals, such as zeolites, carbonates, quartz, or chalcedony, to form what are referred to as *amygdales* (IT 15) (amygdule - diminutive). Rocks containing amygdales are said to be *amygdaloidal*. Amygdales may contain concentric arrangements of minerals which were deposited on the walls of the original vesicle. More rarely these minerals may be deposited in horizontal layers. When found in lavas that have been tilted, the attitude of these layers relative to the attitude of the lava can provide information on the time of deposition of the secondary minerals relative to that of the tilting.

Exsolution of gas from magma can lead to violent eruptions in which the volcanic material is fragmented into particles ranging from dust to large blocks. This material is collectively referred to as *tephra* or *pyroclastic* material (broken by fire). Fragments less than 2 mm in diameter are called *ash*, those between 2 and 64 mm, *lapilli*, and those greater than 64 mm, *blocks* or *bombs*. Upon lithification volcanic ash forms a rock called *tuff*, whereas the coarser materials form *agglomerate* or *volcanic breccia*. These pyroclastic rocks can be categorized by reference to the physical nature of the constituent fragments. *Vitreous tuff*, for example, consists largely of glassy fragments, *crystal tuff* of broken crystals, and *lithic tuff* of rock fragments. The fragments in vitreous tuffs commonly have a characteristic shape which reflects their origin as isolated patches of liquid trapped between coalescing vesicles. These glassy splinters or *shards* are typically bounded by concave surfaces.

Much pyroclastic material cools and solidifies before falling back to earth. Large bombs landing close to their source would be an obvious exception; these may flatten or even burst on landing, extruding their molten interiors. One important class of pyroclastic material, however, remains hot enough during eruption to permit its fragments eventually to weld together into solid rock. Dense, ash-laden clouds that erupt from volcanoes as highly fluid suspensions flow rapidly down slope, coming to rest in topographic depressions. Here particles settle and weld themselves together to form a *welded tuff* or *ignimbrite* (IT 5). The compaction that accompanies the welding flattens the hot glassy fragments and forms a rock with a prominent laminated structure referred to as *eutaxitic*. Because these eruptions involve pyroclastic material that travels in a coherent flow or avalanche, they are referred to as *ash flows* to distinguish them from ash that falls freely through the air, blanketing the ground evenly regardless of local topography.

2. Structures in Plutonic Rocks

The most common structure in plutonic igneous rocks, especially the mafic and ultramafic ones, is *layering*. Layers may result from variations in the abundance or grain size of minerals (IT 41-42), or from elongate minerals being orientated parallel or perpendicular to the layering. Layering may also be caused by variations in the composition of minerals, but because this can be discerned only through chemical analysis (and consequently is not evident in the field) it is referred to as *cryptic layering*. All or any combination of these various forms of layering may be present in a layered rock. Layers also tend to repeat themselves, giving rise to *rhythmic layering* (IT 42).

Layering has been interpreted as resulting from gravitative settling of minerals from magma. This is most convincing where the layers are subhorizontal and exhibit grading similar to that found in some sedimentary rocks where the larger and denser minerals are concentrated towards the base of a layer and the finer and less dense minerals towards the top (IT 42). In some intrusions, the layering may parallel vertical walls, or even the roof, in which case gravitative settling seems an unlikely explanation. Furthermore, some experimental evidence indicates that minerals may not sink, despite appropriate density contrasts, because the crystals lack sufficient force to overcome the yield strength of the magma. In such cases, mineralogical layering might be the result of diffusion processes. That layering can form in such a way is demonstrated in the laboratory by the well-known Liesegang's rings phenomenon.

Layering can also be formed by the flow of magma. This type tends to parallel contacts, and, if feldspar laths are present, the layers may exhibit trachytoidal texture. In many granites, feldspar phenocrysts may be aligned, even though the groundmass may lack obvious flow structure. Separation of minerals may occur through the process of *flowage differentiation*, whereby minerals of different densities and shapes develop different velocities away from a contact as a result of the shearing effect of the flowing magma near the contact. Evidence of flow is clear when previously-formed layers are scoured out and new layers fill erosional troughs in the earlier rocks. The effects of flow can also be seen where inclusions of country rock are streaked out to form *schlieren*.

Another striking example of layering is exhibited by rocks with an *orbicular structure* (IT 54). Orbicules consist of alternating felsic and mafic layers concentrically arranged about some nucleus, which may be a phenocryst or fragment of foreign material. Individual layers have thicknesses of millimeters, and the orbicules are many centimeters in diameter. They are found in a wide range of rock types but are most common in granites. Their growth has been variously interpreted as resulting from rhythmic crystallization of minerals about a nucleus, or from diffusion processes. There may indeed be both kinds. Orbicules with layers of highly variable and irregular thickness are most easily explained as forming through deposition on a nucleus, but those in which the spacing between layers widens in a systematic manner towards the margins could have a diffusion origin.

A peculiar type of layering, known as *harrisitic* or *crescumulate*, consists of crystals of olivine and plagioclase that grow upward from the floor of a magma chamber. The crystals, which may be up to 20 cm long, presumably grow into the magma from nuclei on the floor of the chamber. Hornblende needles with similar crescumulate structure occur in some alkali gabbros, but here the layering tends to parallel the walls of the magma chamber and the hornblende needles are oriented horizontally. A related structure is found in many granitic pegmatites where crystals of alkali feldspar, mica, or tourmaline may nucleate on the wall and grow into the pegmatite perpendicular to the contact. The resulting arrangement of crystals is commonly referred to as a *comb structure*.

Gases dissolved in plutonic magma must be either incorporated into hydrous or other volatile-bearing minerals or be exsolved during crystallization. With slow cooling and solidification the volatiles may be completely expelled from the main igneous body to form pegmatites or hydrothermal veins in the surrounding rock. In some near surface plutonic rocks, however, the volatiles may become trapped and form crystal-lined vugs known as *miarolitic* cavities.

In rare cases, heat liberated by cooling mafic bodies may be sufficient to cause melting of quartz and alkali feldspar in the intruded rock. The resulting granitic liquid, which is referred to as *rheomorphic*, may back vein the rock from which the heat came. *Rheomorphism*, however, is rare because contact temperatures are typically half the initial temperature of a magma, and that is not high enough to melt granite. Contact temperatures can be raised only by the magma continuing to flow, especially in a turbulent manner (only possible for mafic dikes wider than about 5 meters), or having another igneous body intruded nearby. The presence of rheomorphism along dikes can thus be important evidence of feeder flow.

3. **Structures Resulting from Inclusions**

Many igneous rocks contain inclusions of other rocks that appear foreign and consequently are termed *xenoliths*. Xenoliths may be *accidental*, if composed of rock that is completely unrelated to the igneous rock in which they are found, or they may be *cognate*, if formed of rock that is genetically related to the igneous host rock. This distinction is not always easily made. Xenoliths can consist of individual crystals, in which case they are called *xenocrysts*. Some cognate xenoliths are formed by phenocrysts that have clustered and grown together to form the glomeroporphyritic texture.

Xenoliths may be angular and have sharp boundaries, but most show some rounding and diffuseness. This depends on the relative melting points of the xenolith and the surrounding igneous rock and whether the minerals of the xenolith bear a reaction relation to the magma. An accidental xenolith, for example, is more likely to contain minerals that are not in equilibrium with a magma than is a cognate one.

It is not uncommon for magmas of contrasting composition to commingle, in which case rounded bodies of one rock type are enclosed in another. This is particularly true of basaltic and granitic magmas, where the basalt forms pillow-like bodies in the granite. Because of different melting points, densities, and viscosities, these commingled magmas show little, if any, evidence of dissolving in one another. Indeed, the difference in melting points commonly results in the granitic magma quenching the basaltic pillows to relatively fine-grained rock.

An igneous rock containing abundant xenoliths is termed an *igneous breccia*. In the extreme case of *diatreme breccias*, there is no igneous matrix at all; instead, larger fragments are embedded in more finely comminuted material to form a rock, *tuffisite*, which resembles concrete in appearance. Most of the fragments in these breccias are of the surrounding country rock, but some may be derived from the lower crust or upper mantle. These xenoliths, many of which contain minerals indicative of high pressures such as pyrope-rich garnet and jadeitic pyroxene, are typically rounded, so are referred to as *nodules*. Mantle nodules are surrounded by prominent reaction rims which attest to the instability of the high-pressure minerals at the low pressures encountered in the crust. Mantle nodules are also found in some alkaline basalts.

TEXTURES AND STRUCTURES OF IGNEOUS ROCKS

CRYSTAL DEVELOPMENT

1. Euhedral crystals of quartz in welded tuff, Ardnamurchan, Scotland. The dipyramid shape of the crystal indicates that it crystallized as the high-temperature beta form; on cooling, it inverted to the low-temperature alpha form. Plane light, X23.

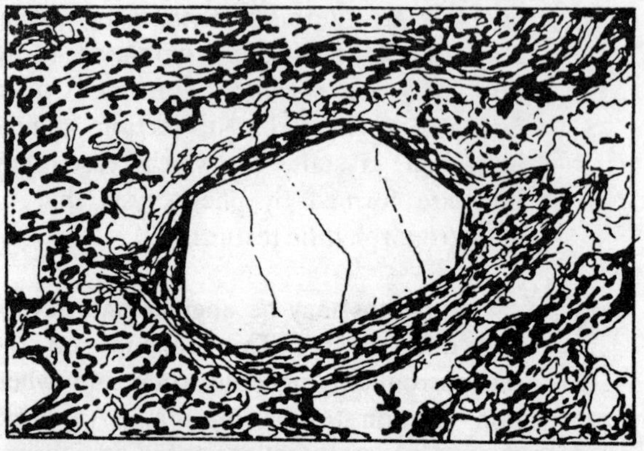

2. Anhedral grains of quartz and perthitic alkali feldspar in quartz mangerite, Grenville, Quebec. Because no grains have crystal faces, the rock is described as being allotriomorphic. Crossed polars, X23.

3. Quartz syenite, South Pond, New Hampshire. Some of the large perthitic feldspar grains have crystal faces, but the quartz grains do not. The rock is therefore described as being hypidiomorphic. Crossed polars, X9.

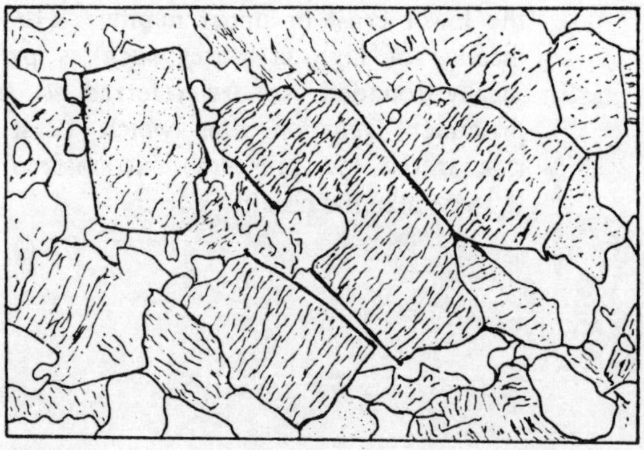

4. Welded tuff consisting largely of volcanic glass enclosing crystals and crystal fragments of quartz and sanidine, Ardnamurchan, Scotland. Cooling of the siliceous magma was rapid enough to quench it to a glass. The crystals of quartz and feldspar were crystallizing prior to eruption. Plane light, X23.

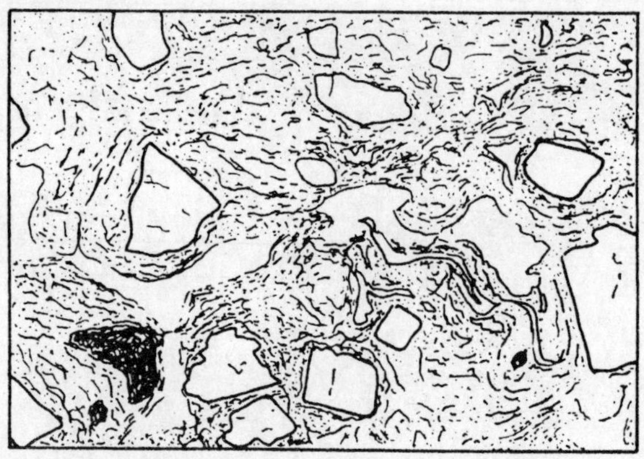

TEXTURES AND STRUCTURES OF IGNEOUS ROCKS

GRAIN SIZE

5. Welded tuff, Castle Rock, Colorado. Small glass shards, which are formed by the explosive disaggregation of vesiculating magma, have been flattened and welded together to give the rock a eutaxitic texture. Plane light, X23.

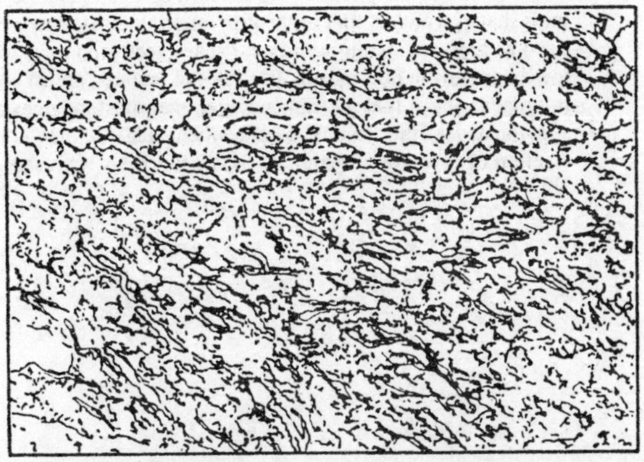

6. Fine-grained Hawaiian basalt with microlites of plagioclase wrapping around an olivine phenocryst. The crystals in the groundmass of this basalt are so fine that they are distinguishable only under the microscope. Crossed polars, X9.

7. Coarse-grained gabbro, Lake St. John anorthosite complex, Quebec. Individual plagioclase laths are up to several centimeters long, and the pyroxene (dark) forms a single crystal in each of these samples, growing around the plagioclase laths in an ophitic texture.

8. Exceptionally large crystal of plagioclase from the Nain anorthosite massif, Labrador. The iridescent color results from scattering of light from compositionally different lamellae within the plagioclase (p. 45) which make a small angle to the (010) plane. The lamellae have a common orientation throughout this boulder, indicating that the boulder is all part of one crystal. Grain sizes such as this are described as pegmatitic.

9. Felsite porphyry with phenocrysts of quartz, Salem, Massachusetts. Although the phenocrysts are now low-temperature quartz, their shape, which is that of the dipyramid, indicates they crystallized as high-temperature quartz. Prism faces predominate in the low-temperature polymorph. Porphyritic textures are so common in igneous rocks that we can conclude that most magmas are not superheated at the time of emplacement; that is, they contain a fraction of crystalline material--the phenocrysts. Crossed polars, X9.

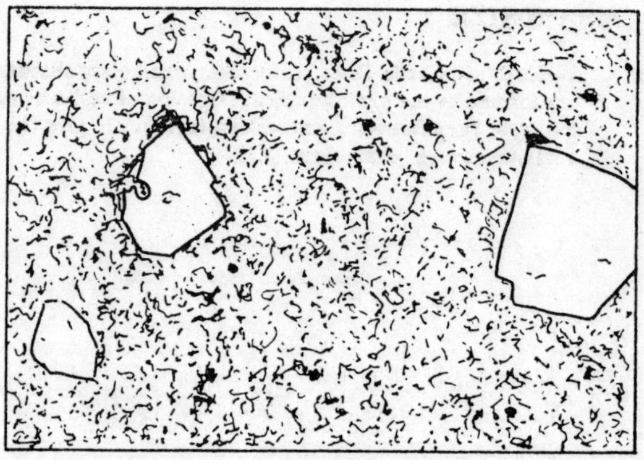

10. Vitrophyric pitchstone, Island of Mull, Scotland. Spherulites of feldspar nucleated on feldspar phenocrysts. Microlites of pyroxene occur throughout the glass, except in the immediate vicinity of pyroxene phenocrysts (lower left) because in these areas the melt did not become supersaturated in pyroxene. Plane light, X23.

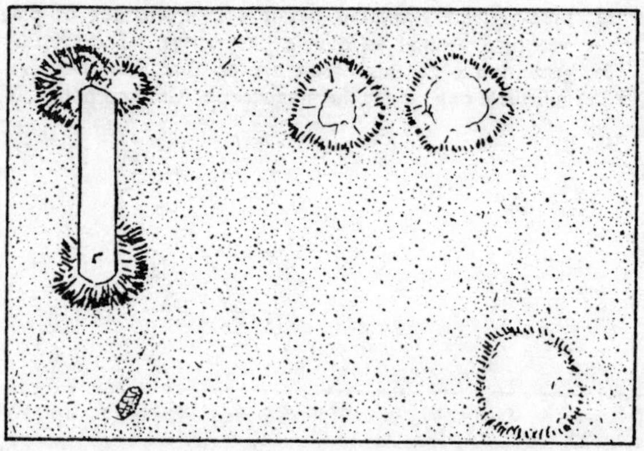

11. Radiating needles of riebeckite around quartz and feldspar crystals in a fine-grained felsite dike, Chatham-Grenville intrusion, Quebec. The quartz and feldspar phenocrysts provided centers on which the riebeckite crystals nucleated. Plane light, X9.

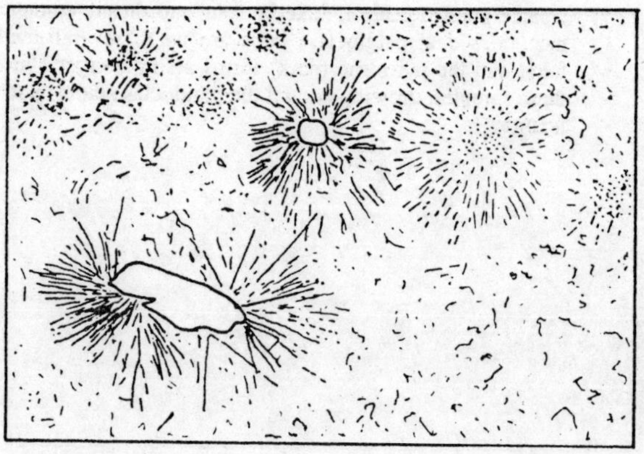

TEXTURES AND STRUCTURES OF IGNEOUS ROCKS

EXSOLUTION OF GAS

12. Vesicular Hawaiian basalt. The spherical shape of the vesicles indicates that gas was still being exsolved when the lava solidified. Had exsolution ceased, the cooling and shrinking gas would have caused the vesicles to collapse and become convoluted. Vesicles in pahoehoe tend to be spherical, but in most aa flows they are convoluted and collapsed. Crossed polars, X9.

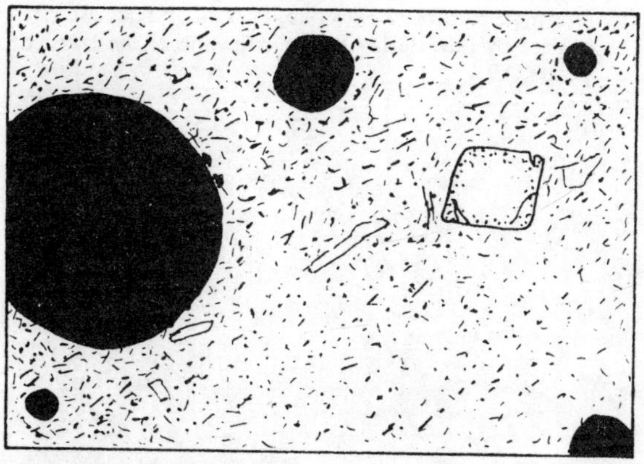

13. Dictytaxitic basalt, Newberry volcano, Oregon. Exsolution of gas in this basalt occurred at such a late stage of solidification that the gas could occupy only the spaces between crystals (dark areas in extinction). Crossed polars, X88.

14. Pipe vesicles formed perpendicular to the walls of a basaltic pillow, Farmington, Connecticut. The crystallizing magma exsolved gas onto bubbles that were imbedded in the advancing crystallization front. As this front moved inward, the bubbles grew as long pipes. Similar pipe vesicles occur at the base of many flows, where they extend upward and commonly are tilted in the direction of flow of the lava.

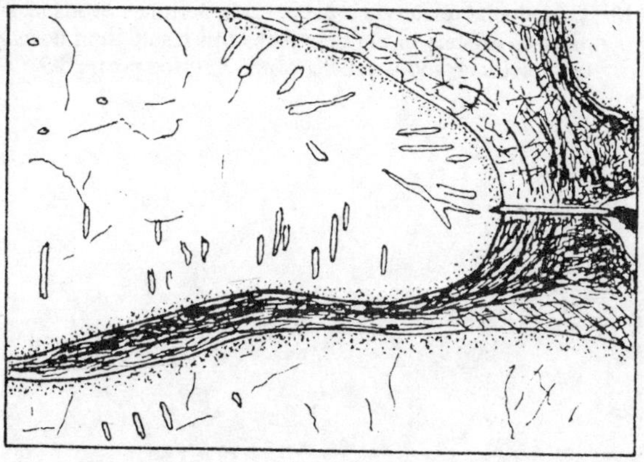

15. Amygdale filled with chlorite (small hemisphere and rim), highly birefringent calcite, and low birefringent zeolite in an alkali basalt, Tahiti. The sequence of deposition in the vesicle was as listed above. Crossed polars, X23)

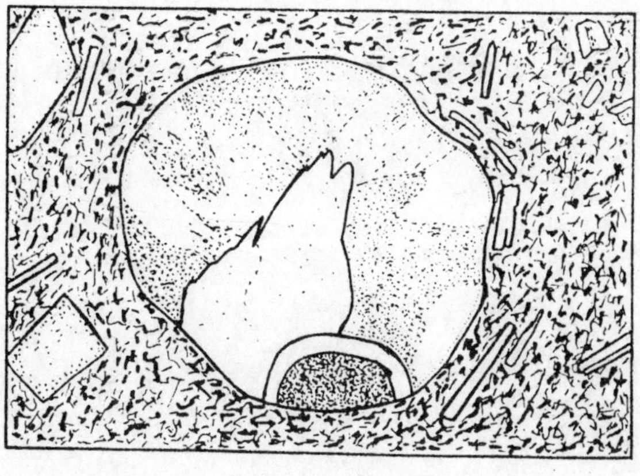

TEXTURES AND STRUCTURES OF IGNEOUS ROCKS

CRYSTAL ALIGNMENT

16. Trachytic texture in Hawaiite, Hawaii. Plagioclase laths were aligned by flow during the solidification of the lava. Crossed polars, X9.

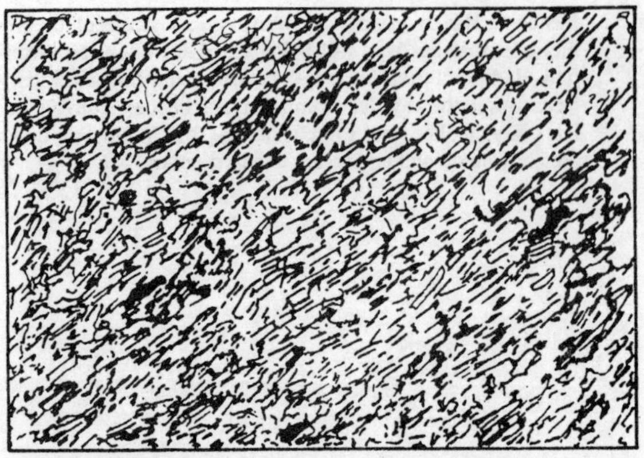

17. Trachytic texture marking flow around a corroded phenocryst of hornblende in a basalt, Haleakala, Hawaii. Crossed polars, X9.

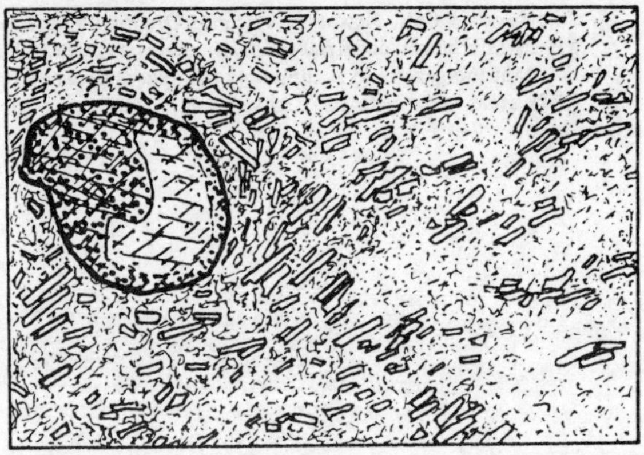

18. Trachytoidal texture in syenite, Loch Borrolan, Scotland. Alignment of crystals in plutonic rocks can result from flow, gravitative settling, and parallel growth. Crossed polars, X9.

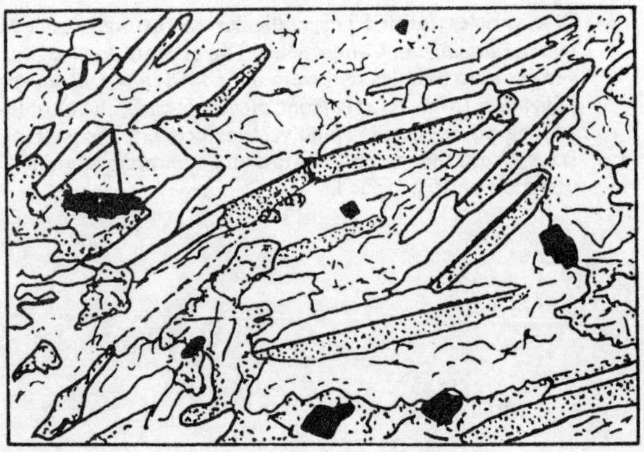

TEXTURES AND STRUCTURES OF IGNEOUS ROCKS

INTERGROWTH TEXTURES

19. Phase diagram for the system Albite--Silica. The eutectic composition of 40% silica and 60% albite closely approaches that of sodic granites, suggesting that these rocks form from minimum melting compositions.

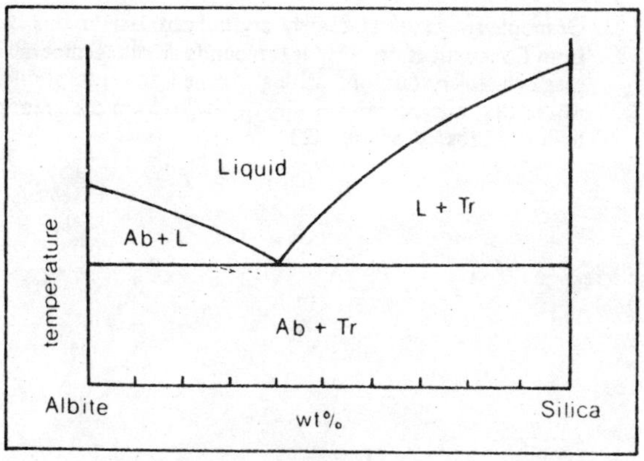

20. Graphic granite. In this section, rods of quartz that pass through the alkali feldspar crystal are seen end on and exhibit the typical shapes that resemble cuneiform writing. This texture may result from eutectic crystallization of these two minerals.

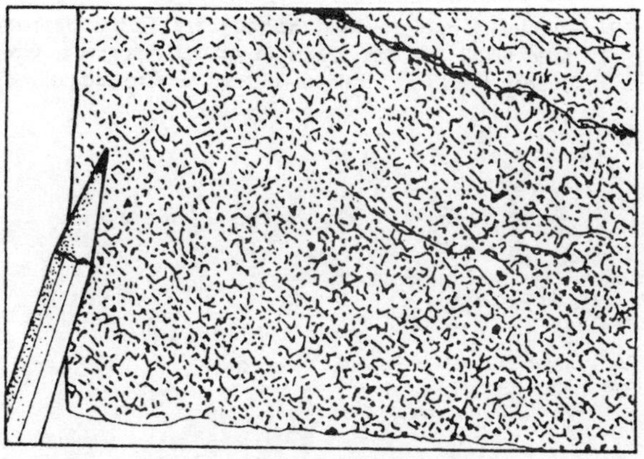

21. Graphic granite with perthitic exsolution lamellae in the host alkali feldspar. From a pegmatite at Portland, Connecticut.

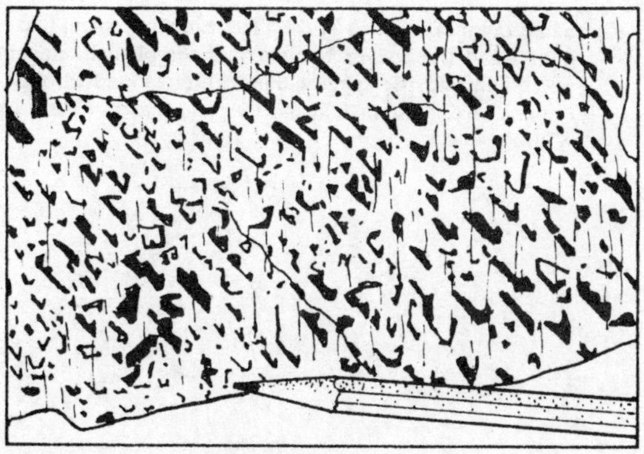

22. Micrographic or granophyric texture in the residual liquid of a Mesozoic diabase dike, Connecticut. Two euhedral plagioclase crystals (in extinction) were crystallizing prior to the coprecipitation of feldspar and quartz. Crossed polars, X54.

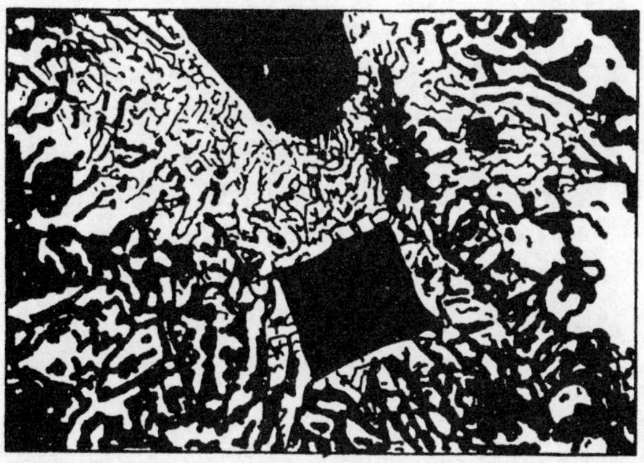

INTERGROWTH TEXTURES

23. Granophyric material clearly crystallized last in this diabase from Connecticut, because it surrounds all other minerals. The plagioclase crystals are strongly zoned towards albitic rims where they are intergrown with quartz to form the granophyric texture. Crossed polars, X23.

24. Myrmekitic intergrowth at the boundary of plagioclase and perthitic alkali feldspar grains in quartz mangerite, Grenville, Quebec. See IT 25 for details of texture. Crossed polars, X23.

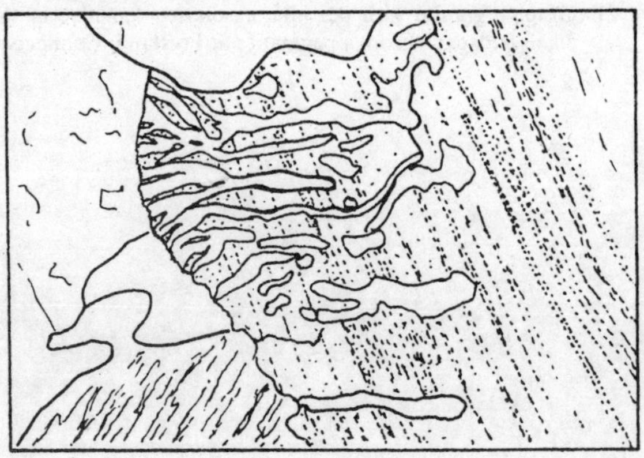

25. Same myrmekite as in IT 24. The radiating worm-like rods of quartz and the convex boundary of the oligoclase against the potassium feldspar suggest this texture formed through replacement of potassium feldspar by plagioclase. Crossed polars, X54.

TEXTURES AND STRUCTURES OF IGNEOUS ROCKS

INTERGROWTH TEXTURES

26. Phase diagram for the system Diopside--Anorthite. Diagrams for many other common ferromagnesian minerals with plagioclase are similar to this one. Many basalts have compositions near the eutectics in these systems.

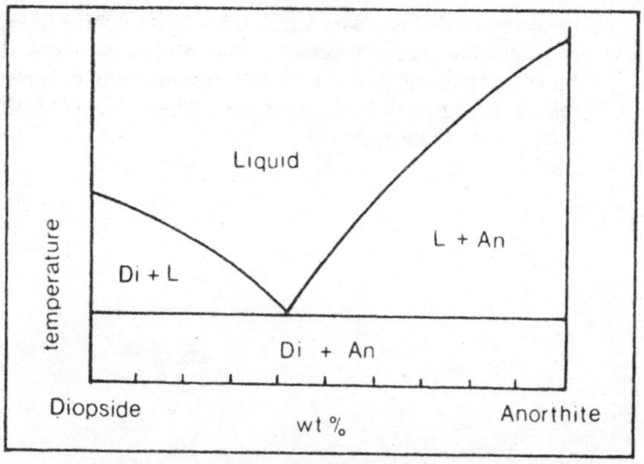

27. Ophitic texture in Karroo dolerite, Birds River, South Africa. This texture probably develops as a eutectic intergrowth. There are many more plagioclase crystals than ferromagnesian crystals in the ophitic texture. As a result, plagioclase laths appear suspended in large ferromagnesian crystals. Crossed polars, X23.

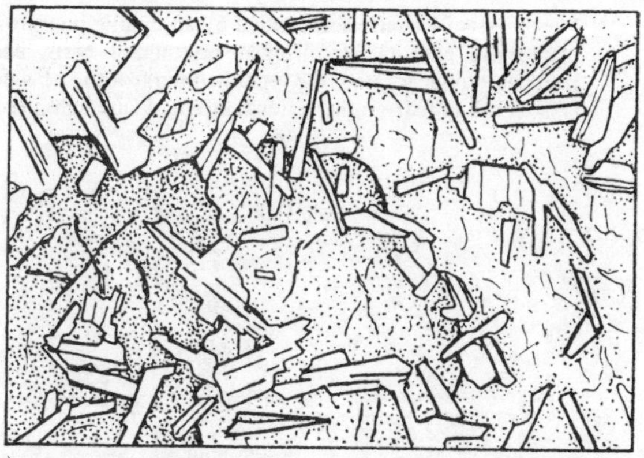

28. Subophitic texture in Mesozoic diabase dike, Cheshire, Connecticut. In this texture there are, on average, equal numbers of plagioclase and ferromagnesian crystals. Consequently, each triangular area between plagioclase laths is occupied by one ferromagnesian crystal. Crossed polars, X88.

29. Intergranular texture in diabase. This sample is from the same dike as that in IT 28, but it comes from closer to the contact where cooling was more rapid. In the intergranular texture, each triangular space between plagioclase laths contains, on average, more than one ferromagnesian crystal. With more rapid cooling still, the residual liquid in this dike crystallized to a fine-grained mesostasis. Crossed polars, X88.

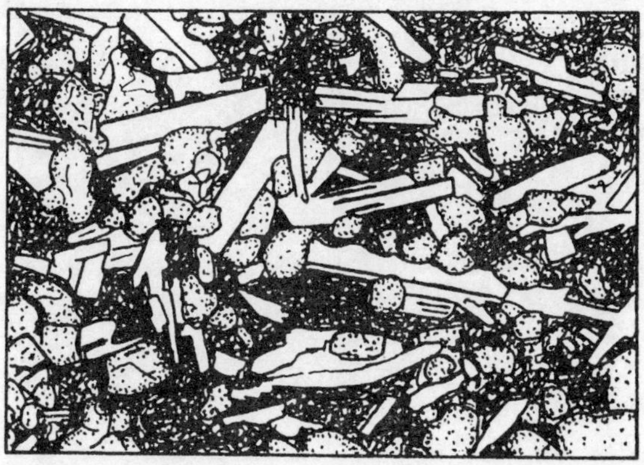

TEXTURES AND STRUCTURES OF IGNEOUS ROCKS

INTERGROWTH TEXTURES

30. Phenocrysts of plagioclase surrounded by an ophitic intergrowth of plagioclase and pyroxene, Ardnamurchan, Scotland. Plagioclase crystallized early, and the residual liquid formed the ophitic intergrowth. Pyroxene crystallized only from the residual liquid. Plane light, X9.

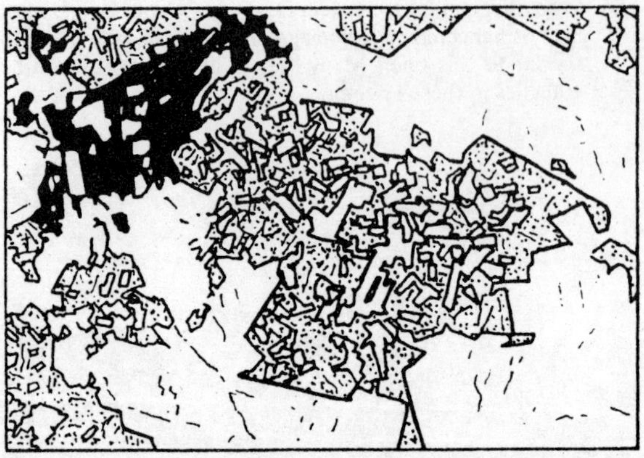

31. Phenocrysts of augite surrounded by an ophitic intergrowth of plagioclase and augite. Augite crystallized early, and the residual liquid formed the ophitic intergrowth. Plagioclase crystallized only from the residual liquid. Plane light, X9.

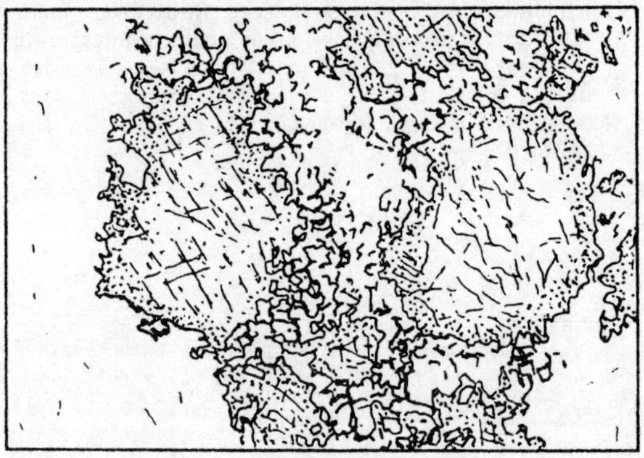

TEXTURES AND STRUCTURES OF IGNEOUS ROCKS

REACTION TEXTURES

32. Phase diagram for the system Forsterite--Silica, showing that enstatite melts incongruently to forsterite and a liquid that is more siliceous than enstatite. Liquids which first crystallize olivine will, on cooling to the incongruent melting point of enstatite (peritectic), react with the olivine to form enstatite.

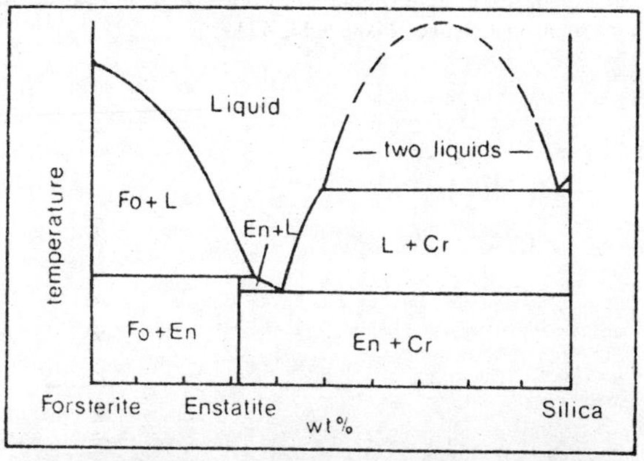

33. Chromite-bearing peridotite from the Stillwater complex, Montana. The liquid which surrounded the early crystallizing grains of olivine and chromite reacted with the olivine to form orthopyroxene. Prior to reaction, the olivine and chromite grains were in contact. The enclosing of numerous grains of olivine and chromite within large crystals of orthopyroxene gives this peridotite a poikilitic texture. Crossed polars, X9.

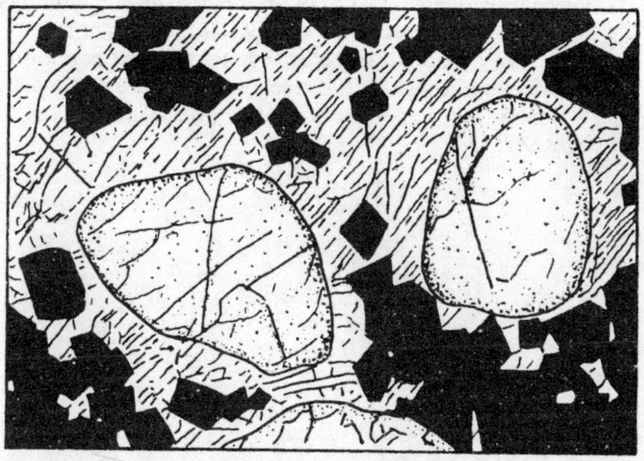

34. Corona texture with augite rimming olivine, Stillwater complex, Montana. Crossed polars, X23.

35. Corona texture with augite rimming orthopyroxene, Bushveld complex, South Africa. Crossed polars, X9.

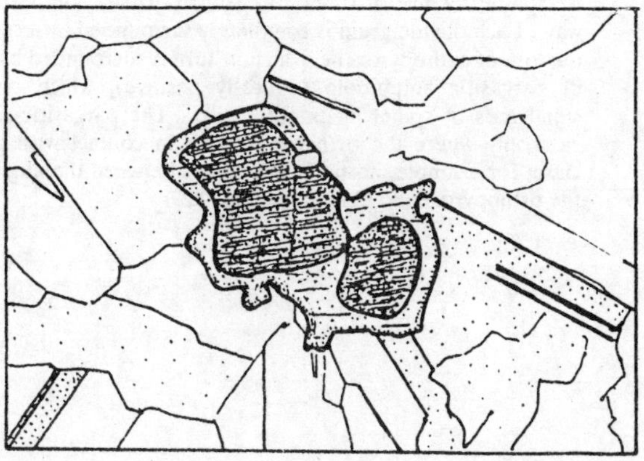

TEXTURES AND STRUCTURES OF IGNEOUS ROCKS

REACTION TEXTURES

36. Augite rimmed and replaced by hornblende in essexite, Mount Johnson, Quebec. Plane light, X23.

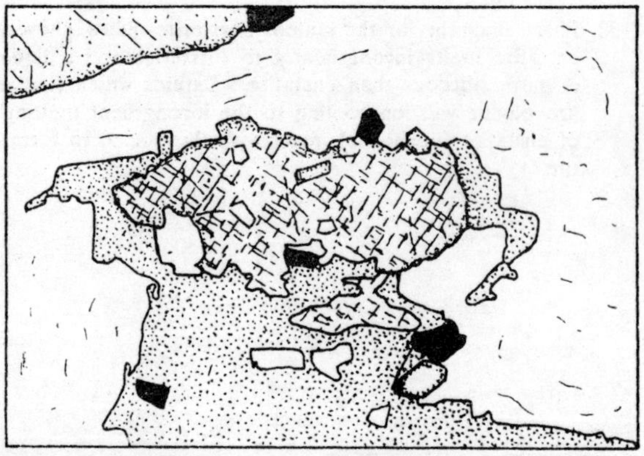

37. Olivine surrounded by rim of orthopyroxene in hybrid rock formed at the contact of an alkaline peridotite, Mount Bruno, Quebec. Assimilation of siliceous sediments by the peridotite magma was responsible for the reaction to produce ortho-pyroxene, a mineral that is not normally found in alkaline rocks. Plane light, X9.

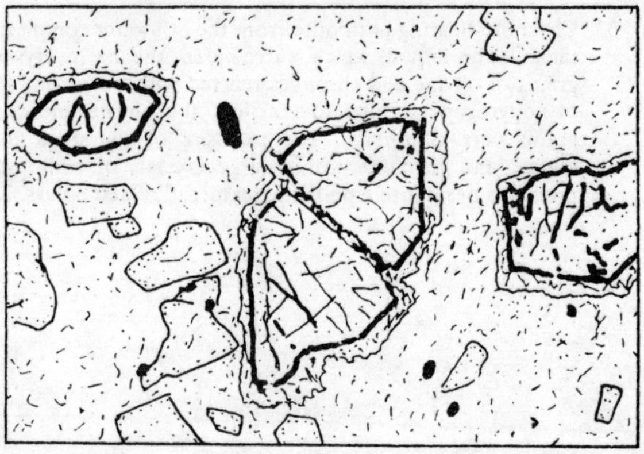

38. Titanaugite crystal in the same hybrid rock as in IT 37 reacted with the contaminated magma to produce a corona of ortho-pyroxene and biotite. Plane light, X9.

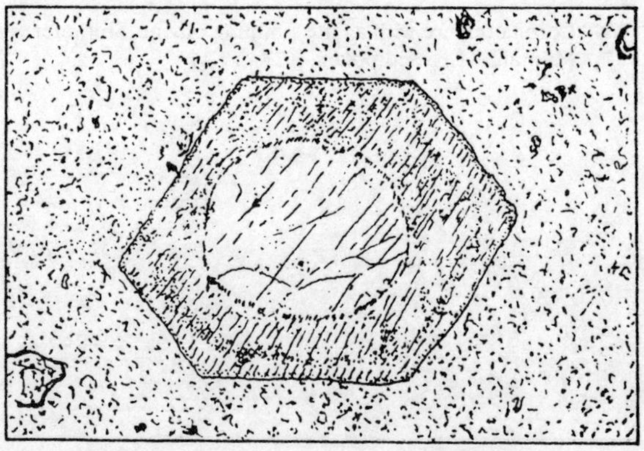

39. Kelyphitic texture in troctolitic gabbro, Risor, Southern Norway. Each olivine grain is completely surrounded by a continuous rim of orthopyroxene, which in turn is surrounded by a rim of pargasitic amphibole (optically positive), which contains small rods of spinel in its outer part. The pargasitic rim occurs only where the orthopyroxene is in contact with plagioclase; for example, no pargasite occurs between the augite and the orthopyroxene. Crossed polars, X29.

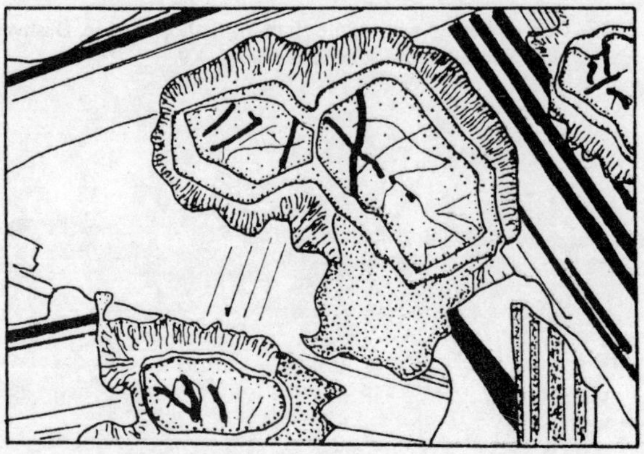

TEXTURES AND STRUCTURES OF IGNEOUS ROCKS

CRYSTAL ACCUMULATION

40. Sinking of olivine crystals toward the bottom of a platinum crucible in which a forsterite--diopside liquid was held a few degrees below the olivine liquidus. (From Bowen, 1915; reproduced with permission of the American Journal of Science)

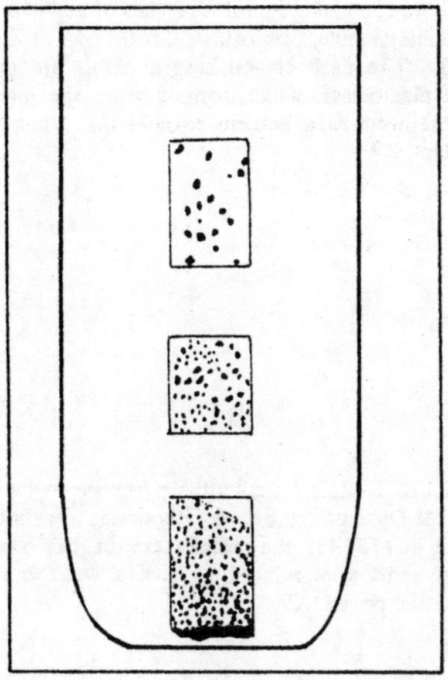

41. Layers of crystal accumulation in the Kiglapait intrusion, Labrador. Note small troughs filled mainly with olivine and pyroxene (dark) in several of the layers.

42. Close-up of outcrop in IT 41. Each layer exhibits grading from an olivine-rich base, through a pyroxene-rich central part, to a plagioclase-rich top. Such graded layering has been interpreted as resulting from gravitative settling of minerals of different mass from dense currents of crystal-rich magma in the same way that sedimentary graded bedding is formed from turbidity currents.

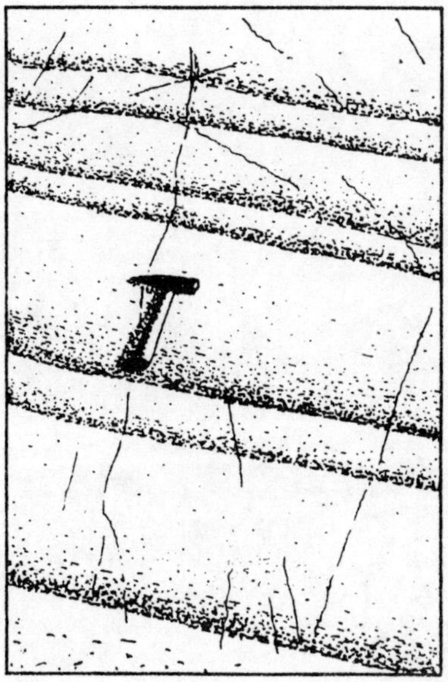

CRYSTAL ACCUMULATION

43. Gabbro containing cumulus crystals of augite and orthopyroxene and intercumulus plagioclase, Great Dyke of Zimbabwe (Rhodesia). The early-crystallizing minerals are euhedral, whereas the plagioclase, which formed from the intercumulus liquid, crystallized as a cement around the other grains. Crossed polars, X9.

44. Gabbro containing cumulus pyroxene and plagioclase grains, Great Dyke of Zimbabwe (Rhodesia). In contrast with the texture in IT 43, the plagioclase in this sample crystallized early and accumulated to give a weak horizontal foliation. Crossed polars, X9.

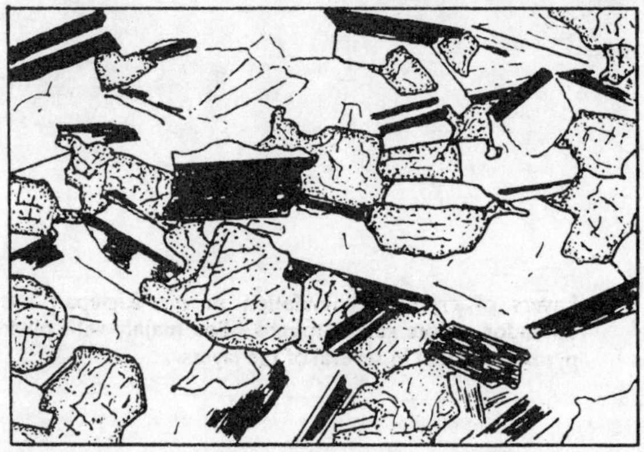

45. Intercumulus liquid may crystallize in situ between cumulus grains (plagioclase in this example), or be expelled by compaction and diffusion from the interstices. Additional growth on the cumulus grains can give rise to an adcumulate, which in extreme cases can produce a monomineralic rock.

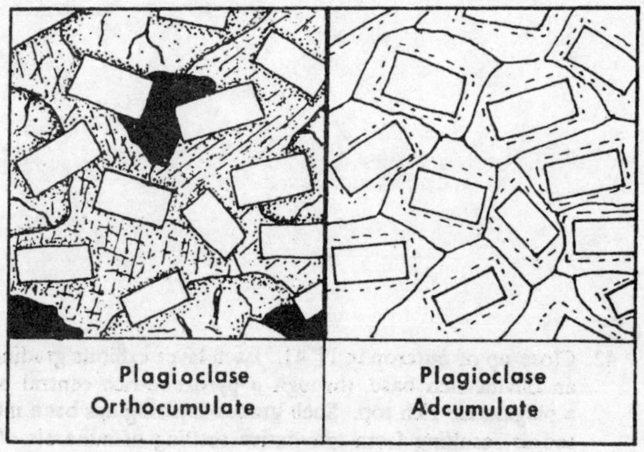

TEXTURES AND STRUCTURES OF IGNEOUS ROCKS

HYPERSOLVUS AND SUBSOLVUS CRYSTALLIZATION

46. Phase diagram for the alkali feldspar system at low and high pressures of water. With increased pressure, melts are able to dissolve more water, which in turn lowers melting points. Above a critical water content (and pressure) the alkali feldspar solvus intersects the solidus, and the liquidus minimum changes to a eutectic. Above this critical value, two different feldspars crystallize from the melt, whereas below it only a single alkali feldspar crystallizes from the melt.

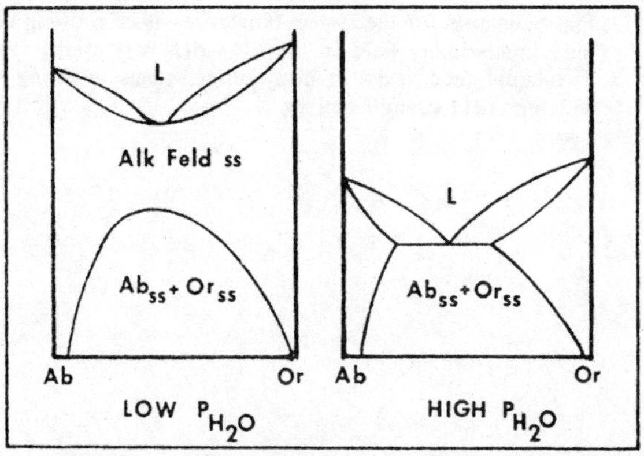

47. Hypersolvus granite, Rockport, Massachusetts. Only one feldspar crystallized from the magma; on cooling this feldspar exsolved to form perthite. We can conclude that this rock formed from a relatively dry magma. Crossed polars, X9.

48. Subsolvus granite, in which two different feldspars crystallized from the melt, a microcline and an albite; Concord, New Hampshire. We can conclude that this granite formed from a relatively water-rich magma. Crossed polars, X23.

TEXTURES AND STRUCTURES OF IGNEOUS ROCKS

LIQUID IMMISCIBILITY

49. Phase diagram for the system Forsterite--Silica showing the liquid immiscibility field in the silica-rich part of the system. Two-liquid fields exist in many other sytems involving silica with high field strength cations.

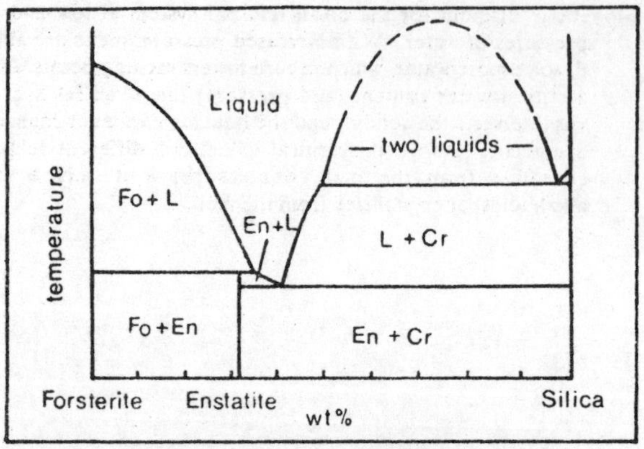

50. Silicate liquid immiscibility in the residual liquid of a tholeiitic basalt, Kilauea, Hawaii. The two liquids are preserved as droplets of brown iron-rich and clear silica-rich glasses. Plane light, X480.

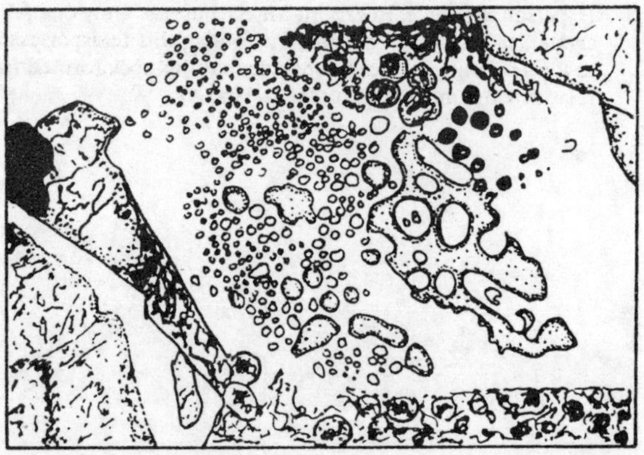

51. Immiscible droplets, especially when crystallized, are more readily visible in reflected light. in this tholeiitic basalt from Connecticut, the iron-rich droplets appear as reflective spheres, speckled with minute crystals of magnetite, in a nonreflective silica-rich glass between laths of plagioclase and pyroxene. Dendritic magnetite crystals (bottom center) are common in the mesostasis of basalts exhibiting immiscibility. Reflected plane light, oil immersion, X350.

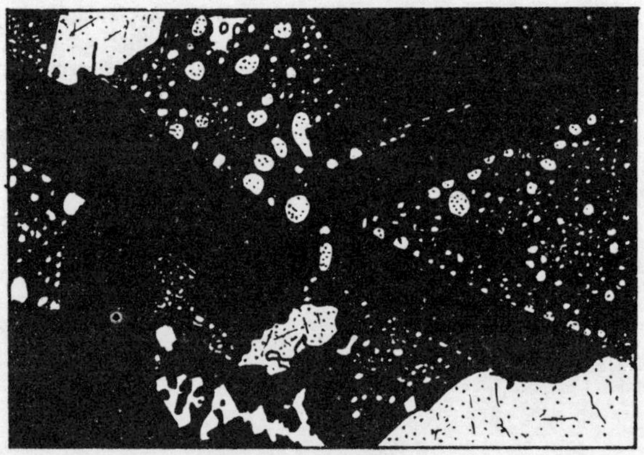

52. Mixed basaltic and rhyolitic glasses in a pumice, Newberry volcano, Oregon. The juxtaposition of these two different liquids is a result of commingling of two magmas and not liquid immiscibility. Note the lack of small droplets of one liquid in the other which are formed when immiscibility is involved (IT 51). Plane light, X11.

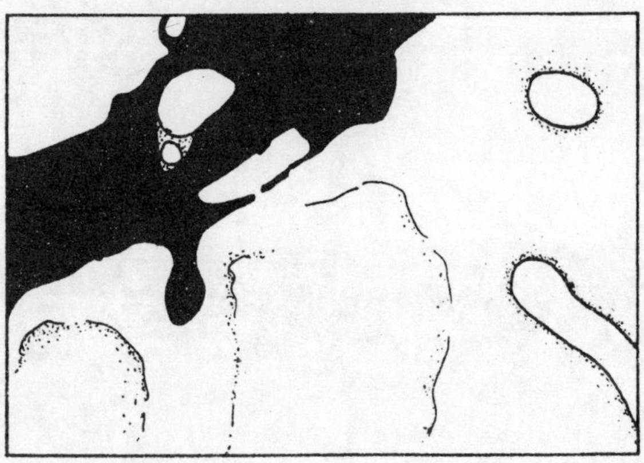

TEXTURES AND STRUCTURES OF IGNEOUS ROCKS

LIQUID IMMISCIBILITY

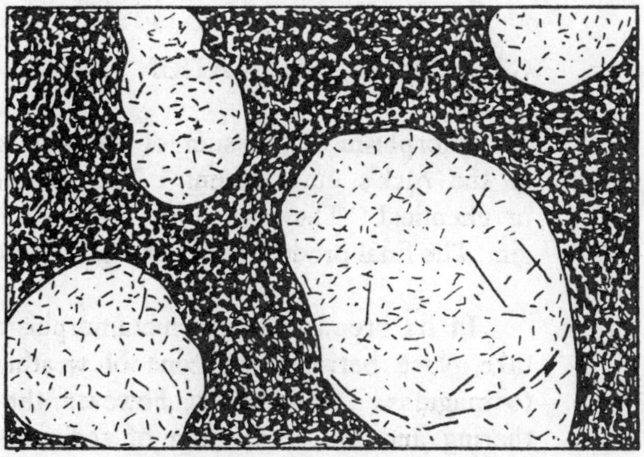

53. Ocelli in lamprophyric dike, Montreal, Quebec. These segregations of felsic materials have been interpreted both as globules of immiscible liquid and as the fillings of vesicles with late-stage magmatic liquid. Plane light, X9.

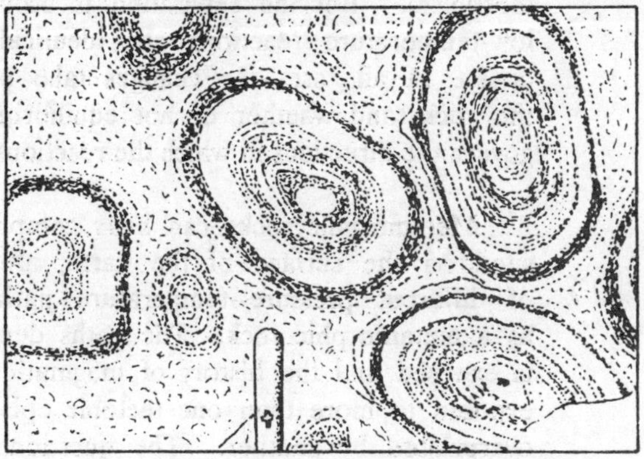

54. Orbicular granite, Kangasala, Finland. Although once believed to form through liquid immiscibility, orbicules are the product of rhythmic precipitation of felsic and mafic minerals about some nucleus, which is commonly a phenocryst or xenolith. Swiss Army Knife for scale.

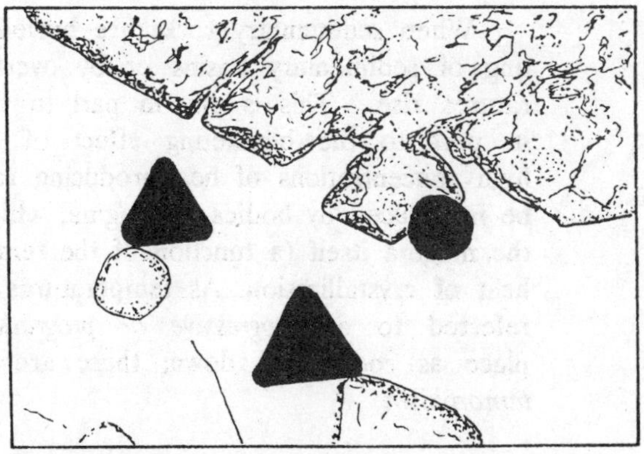

55. Immiscible sulfide droplet in the residual silicate glass of a tholeiitic basalt, Southbury, Connecticut. Plane light, X830.

56. Same as IT 55 but under reflected light, where the sulfide appears yellow compared with the two magnetite crystals which are white.

57. Immiscible sulfide globules in gabbro, Sudbury, Ontario. The outline of the globules is marked by silicate crystals which crystallized first. Coalescence of these immiscible globules played a role in concentrating nickel, which preferentially enters the sulfide liquid, into ore deposits.

6 Classification and Textures of Metamorphic Rocks

GENERAL CLASSIFICATION

Metamorphism is the sum of all changes that occur in a rock in response to changes in the rock's environment. The changes in the rock can be physical, mineralogical, or chemical. They can affect rocks of igneous, sedimentary, or even metamorphic origin. The final product of the changes is known as a metamorphic rock.

In its broadest sense, metamorphism includes the entire spectrum of changes that take place between the zone of weathering and the zone in which melting gives rise to magmas. Traditionally, however, the low-temperature changes associated with weathering and the lithification of sediments have been omitted from the study of metamorphism. But the separation is arbitrary, for the fundamental laws governing the low-temperature reactions are identical to those controlling the higher-temperature ones. In all cases the changes take place in order to minimize the free energy of the rock, and whether or not equilibrium is achieved, basic thermodynamic principles govern the directions in which the reactions proceed.

Metamorphic rocks can have complicated histories. For instance, sediment deposited on the surface of the earth, after becoming lithified, can follow any number of different pressure--temperature paths before reappearing on the earth's surface as a metamorphic rock. The paths depend on the tectonic process to which the rock is exposed and the history of magmatic events in the region. The rock may even be exposed to more than one tectonic episode, in which case one set of changes may be overprinted by another. The question, then, arises as to what part of the cycle of change is actually preserved and recorded in the final metamorphic rock.

When sedimentary rocks are buried, either through continued subsidence and filling of sedimentary basins or by overthrusting of other crustal rocks, their temperatures rise. This is due in part to the geothermal gradient (10 to $60^{o}C/km$) and in part to the blanketing effect of crustal rocks which normally have relatively high concentrations of heat-producing radioactive elements (U, Th, K). Heat can also be introduced by bodies of magma, which on cooling liberate both heat brought in by the magma itself (a function of the temperature and heat capacity of the magma) and heat of crystallization. As temperatures rise, changes take place which are commonly referred to as *progressive or prograde metamorphism*. Reactions can also take place as rocks cool down; these are referred to as *retrogressive or retrograde metamorphism*.

Retrograde metamorphism is far less common than prograde metamorphism. Indeed, most metamorphic rocks are believed to preserve mineral assemblages formed at peak temperatures. There is both a kinetic and chemical explanation for the scarcity of retrograde rocks. The thermodynamic driving potential for most metamorphic reactions varies linearly with temperature. Reaction rates, however, vary exponentially with temperature. During prograde metamorphism, the driving potential and the kinetic factor act in unison to cause reactions to proceed. But with falling temperature, the driving potential is opposed by the exponentially decreasing reaction rate. Furthermore, most prograde reactions involve dehydration or decarbonation. Once volatiles have been expelled from a rock, falling temperatures cannot then reverse these reactions. Consequently, most retrograde metamorphism is a local phenomenon, occurring mainly where fractures provide avenues for fluids to reenter the rock. These fluids are not only reactants, but they also act as catalysts by significantly increasing reaction rates.

The history of a metamorphic rock is recorded in its mineralogy and textures (and to a lesser extent, chemistry). For this reason, mineralogy and texture form the basis for classification. A metamorphic rock is described and classified using mainly descriptive terms. This involves a listing of the constituent rock-forming minerals followed by a textural term. For example, a quartz-muscovite-biotite schist consists predominantly of quartz, muscovite, and biotite, and its mica crystals have a parallel alignment, which imparts a foliation or schistosity to the rock. (A listing of metamorphic textures is given at the end of this chapter.) A few special names are used for metamorphic rocks of striking chemical composition. For example, marble, calc-silicate, quartzite, and soapstone (serpentine and talc) are formed from the metamorphism of limestone, impure limestone, quartz-rich sandstone, and ultramafic rock, respectively. Although these terms are genetic, the distinctive composition of each rock leaves little doubt as to its origin.

The mineral assemblage constituting a metamorphic rock can be used to interpret the temperatures, pressures, and, in some cases, the composition of the fluid phase extant during metamorphism. Furthermore, the textures of the rock can reveal the nature of the disturbance which caused the metamorphism. For example, they might indicate whether the disturbance was a result of deep burial and deformation related to mountain building (*regional metamorphism*), or was a result of the nearby emplacement of an igneous intrusion (*contact metamorphism*). Interpretation of the conditions and causes of metamorphism leads to another way of classifying metamorphic rocks known as the facies classification. Before dealing with this approach, however, it is necessary to discuss the graphical representation of metamorphic mineral assemblages.

GRAPHICAL REPRESENTATION OF METAMORPHIC MINERAL ASSEMBLAGES

One of the important early discoveries of petrology was that metamorphic rocks contain relatively small numbers of minerals. This was correctly interpreted in 1912 by V. M. Goldschmidt as indicating that the minerals in metamorphic rocks closely approach equilibrium assemblages. That is, if a reaction between minerals can take place, it proceeds until one of the reactants is eliminated. In this way, the number of minerals is kept to a minimum. Indeed, Goldschmidt showed, through thermodynamic arguments, that the number of minerals in a rock should normally equal the number of components present, where components are defined as the smallest number of chemical building units (single oxides or groups of oxides) necessary to describe all of the minerals present. Many rocks contain no more than four or five components, and thus they contain only four or five major minerals.

Because of the small numbers of minerals present in most metamorphic rocks, it is possible to construct graphical representations of the mineral assemblages, which aid not only in making petrographic descriptions, but in interpreting the conditions of metamorphism. In three dimensions, four compositional components can be represented, for example, in a tetrahedral plot. Unfortunately, when such a plot is projected into the two dimensions of a sheet of paper, many of its features can become obscured. In two dimensions we are restricted to plotting only three components, which can be represented with a simple triangular graph. Most rocks, however, contain one or two more components than can be represented in the triangular plot. Consequently, the various graphical representations used in studying metamorphic mineral assemblages are means of projecting data from four- or five-component space into the three components of the triangular plot.

One of the most commonly used projections is the ACF diagram, where A, C, and F are defined in mole percent as:

$$A = Al_2O_3 + Fe_2O_3 - (Na_2O + K_2O)$$
$$C = CaO$$
$$F = FeO + MgO + MnO$$

The A value plotted in this diagram includes all of the alumina in the rock minus that which enters alkali feldspar. In this way we eliminate the need to plot the alkalis in the diagram, reducing by one the number of components to be represented. A further reduction is achieved by grouping together FeO, MgO, and MnO. In many metamorphic minerals these oxides freely substitute for one another and, consequently, can be treated as a single component. But in some minerals, such as staurolite, this unfortunately is not true, and this grouping is invalid. It will be noted that the most abundant component of all, SiO_2, is omitted from this plot. It is not, however, forgotten. The silica forms a fourth component which would require a tetrahedral plot if it were to be graphically represented. A point representing the bulk composition of a rock in such a tetrahedron can be projected from any apex of the tetrahedron onto the opposite triangular face (Fig. 6-1). Because most metamorphic rocks contain free silica (in the form of quartz), we can project the bulk composition of the rock from the silica apex onto the ACF face of the tetrahedron. We can then use the triangular ACF diagram to show the composition of the rock and its constituent minerals (in terms of the ACF components), as long as we state that quartz is also present (although not graphically represented in the ACF diagram). Within the limits of the above restrictions a rock can be plotted in the simple ACF triangular diagram.

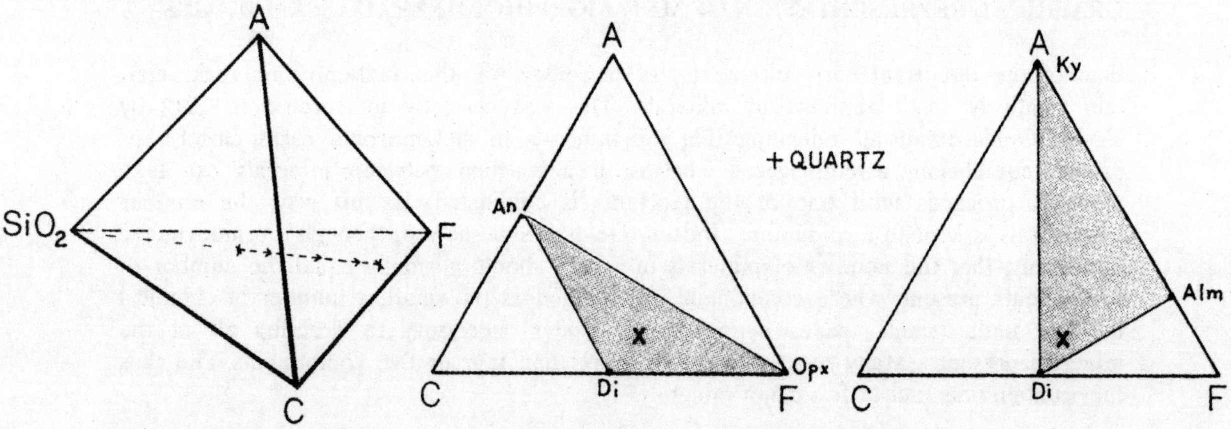

Figure 6-1 ACF diagram for rocks containing quartz. Rock compositions are projected from SiO_2 onto the ACF face of the tetrahedron (see text for explanation). Under one set of metamorphic conditions a rock with bulk composition X might consist of the mineral assemblage quartz-anorthite- diopside-orthopyroxene, but under another set of conditions would consist of quartz-kyanite-diopside-almandine.

To plot the composition of a rock in the ACF diagram, the chemical analysis, which is normally presented in terms of weight percent, is recast in mole proportions by dividing the weight percentage of each oxide by its molecular weight. The mole proportions of the A, C, and F components are then recalculated to 100%. If accurate modal data are available, corrections can be made to the mole proportion of CaO for any accessory apatite or sphene present in the rock, and to FeO for accessory ilmenite and magnetite (see norm calculation, Chapter 4, for calculation). Normally the amounts of these minerals are small, and they do not seriously affect the ACF plot.

The minerals in the rock are also plotted in the ACF diagram. Their positions can be deduced easily from their formulae given in Chapter 3. For convenience, their compositions in terms of A, C, and F are given below:

Mineral	A	C	F
Anthophyllite, Cummingtonite, Orthopyroxene			
Talc, Serpentine	0	0	100
Actinolite, Tremolite	0	28.5	71.5
Hornblende	<25	~28.5 :	71.5
Diopside, Dolomite	0	50	50
Calcite, Wollastonite	0	100	0
Grossularite, Andradite	25	75	0
Vesuvianite	14	72	14
Epidotes	43	57	0
Anorthite	50	50	0
Al_2SiO_5, Pyrophyllite	100	0	0
Staurolite	67	0	33
Chloritoid (Fe), Cordierite (Mg)	50	0	50
Spesartine, Almandine, Pyrope	25	0	75
Chlorite	10-35	0	90-65

Most rocks contain three of the minerals in the ACF diagram, plus quartz (accessory minerals are neglected). Lines are drawn between these minerals to show that they occur together and were probably formed as an equilibrium assemblage under the particular conditions of metamorphism represented by the rock. The bulk composition of the rock plots within the mineral phase triangle. Under different metamorphic conditions, this same bulk composition could produce a different assemblage of minerals. Thus, although the point representing the bulk composition in the ACF diagram would plot at the same place, the enclosing mineral phase triangle would be different (Fig. 6-1).

The ACF diagram is well suited for representing metamorphic rocks formed from most igneous rocks, especially those of basaltic composition, and metamorphosed limestones, dolostones, and impure carbonate rocks. It is not suitable for rocks formed from aluminous sediments, the so-called pelites. When metamorphosed, these rocks typically contain abundant muscovite, except at very high temperatures when they contain potassium feldspar instead. They also contain a number of ferromagnesian minerals, such as staurolite, chloritoid, garnet, cordierite, and chlorite, which are either iron-rich or magnesium-rich; that is, iron and magnesium do not substitute freely for each other in these minerals and therefore cannot be treated as one component as is done in the ACF diagram. If these minerals are plotted in the ACF diagram, assemblages of four rather than three minerals commonly result. For such rocks a completely different plot is used.

Pelitic rocks are composed of five essential oxides, SiO_2, Al_2O_3, MgO, FeO, and K_2O. Because of the limited substitution of FeO for MgO in many pelitic ferromagnesian minerals, all five oxides must be treated as separate components for the purposes of graphical representation. If we plot only those rocks containing excess silica (that is, ones containing free quartz), silica need not be plotted. This, in reality, is simply taking compositions in the five-component system and projecting them from quartz into the four-component system. Very few rocks are eliminated by this restriction, because only rarely do pelitic rocks not contain excess silica. By removing the need to plot silica, the number of essential components needed to represent the composition of pelitic rocks and their constituent minerals is four. This, then, can be done with a tetrahedral plot.

In Fig. 6-2, the essential minerals in silica-saturated pelitic rocks are plotted in terms of Al_2O_3, FeO, MgO, and K_2O. By joining with straight lines those minerals that occur in a rock, a tetrahedral volume is formed with the constituent minerals lying at its corners. For example, we could draw the lines for a rock containing the assemblage muscovite, almandine, staurolite, and kyanite (plus quartz), or one containing muscovite, staurolite, chlorite, and biotite (+quartz). The chemical composition of a rock plots within the tetrahedral volume outlined by the mineral assemblage.

Despite its simplicity, the tetrahedral plot would obviously become difficult to read if a number of mineral assemblages were to be plotted in it. We therefore need a means of simplifying the diagram. First, it will be noted that except for the micas and potassium feldspar, all of the minerals plot on the AL_2O_3-FeO-MgO (AFM) face of the tetrahedron. Furthermore, most pelitic rocks contain muscovite rather than potassium feldspar, except at the highest temperatures of metamorphism where muscovite breaks down. Thus, any muscovite-bearing assemblage with a bulk composition plotting within the tetrahedron can be projected from muscovite onto the AFM face of the tetrahedron, producing a triangular plot (Fig. 6-2). Of course, we

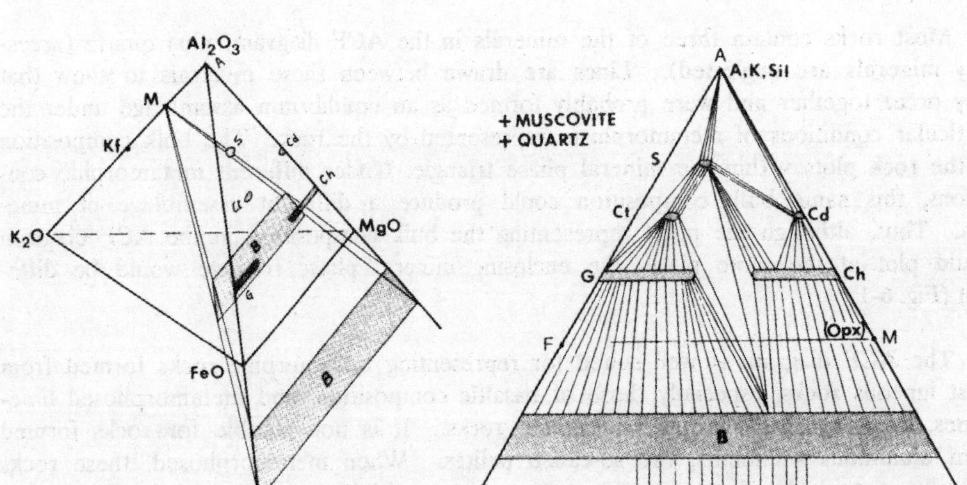

Figure 6-2 Projection of quartz- and muscovite-bearing pelitic rocks onto the AFM face of the Al_2O_3-FeO-MgO-K_2O tetrahedron (see text for explanation). Minerals plotted are Al_2SiO_5 polymorphs (A), staurolite (S), chloritoid (Ct), cordierite (Cd), garnet (G), chlorite (Ch), biotite (B), muscovite (M), and potassium feldspar (Kf). (After Thompson, 1957)

must remember that in addition to the three-phase assemblage in this triangle the rock will also contain muscovite and quartz. For high-temperature pelites, where muscovite is no longer stable, the projection onto the AFM face can be made from potassium feldspar. Because this graphical method was proposed by Thompson (1957) it is commonly referred to as the Thompson AFM projection.

Because most minerals in pelitic rocks plot on the AFM face of the Al_2O_3-FeO-MgO-K_2O tetrahedron, their position on the triangular AFM plot is unaffected by projecting from muscovite. Biotite, however lies within the tetrahedron, so that when projected from muscovite it lies below the line joining FeO and MgO on the AFM plot. In addition, rocks containing muscovite, biotite, and potassium feldspar will project onto the AFM face below biotite.

Substitution of FeO for MgO occurs in most ferromagnesian minerals in meta-pelites, but under any particular set of metamorphic conditions the range of substitution may be rather limited. This range is indicated in the AFM plot by the shaded areas. The degree of substitution of FeO for MgO is greater in some minerals than in others. Where analytical data on the individual minerals are available, for example from electron microprobe analyses, lines known as *tie lines* can be drawn joining the compositions of coexisting minerals. From such data we can generalize that the following minerals show a decreasing preference for magnesium relative to iron: cordierite, chlorite, biotite, chloritoid, staurolite, and almandine. Thus, for example, biotite in a rock will have a higher Mg/Fe ratio than coexisting almandine (Fig. 6-2). Because of the solid solution in ferromagnesian minerals, many pelitic rocks plot in the AFM diagram as an assemblage of only two minerals rather than of three (plus muscovite and quartz); they occupy areas ruled with tie lines joining the compositions of coexisting phases. And with still other rocks, one-phase assemblages are possible in the AFM diagram if the composition of the rock plots in an area occupied by the solid solution range of an individual mineral; the composition of the mineral is then determined entirely by the bulk composition of the rock.

CLASSIFICATIONS REFLECTING CONDITIONS OF METAMORPHISM

The earliest attempt to classify metamorphic rocks on a regional scale was by Barrow (1893) working in the southeastern Scottish Highlands. He was able to map *zones* of progressively metamorphosed rock which he named after readily identifiable minerals; these were chlorite, biotite, garnet, staurolite, kyanite, and sillimanite, listed in order of increasing intensity of metamorphism. The outer part of a zone, where the index mineral makes its first appearance, was named an *isograd*, the implication being that an isograd marked a line of equal grade, or degree, of metamorphism. Mapping of isograds is still used as a convenient field method for studying metamorphic rocks and for giving a general idea of the intensity of metamorphism. But the relation between the position of an isograd and the intensity of the factors causing metamorphism is now recognized to be far from simple. Furthermore, a single index mineral is a far less sensitive gauge of metamorphic grade than the mineral assemblage of the rock as a whole. Thus, for detailed work, modern classifications make use of the complete mineral assemblage.

Eskola, working on the contact metamorphic rocks of the Orijarvi region of Finland, was the first to fully appreciate the broad relations between mineral assemblages, rock composition, and the pressures and temperatures of metamorphism. He noted that within regions that were small enough to have experienced the same grade of metamorphism throughout, mineral assemblages were determined only by rock composition, but that rocks from different areas, covering similar ranges of composition, could contain very different mineral assemblages, which he attributed to different conditions of metamorphism in the different areas. Eskola incorporated these observations into the metamorphic facies concept. He defined a *metamorphic mineral facies* as comprising all the rocks that have originated under temperature and pressure conditions so similar that a definite chemical composition results in the same set of minerals (Eskola, 1920). Thus, for rocks metamorphosed under the same conditions, different mineral assemblages represent different bulk compositions. Conversely, rocks with the same bulk composition but with different mineral assemblages reflect different conditions of metamorphism.

Eleven different facies are recognized, covering the entire spectrum of possible metamorphic conditions. As a result of experimental studies, the approximate pressures and temperatures of the boundaries between the various facies are now known (Fig. 6-3). It should be emphasized that these are only approximate, and that they can be modified considerably by changing the composition of the fluid phase. In Fig. 6-3, the partial pressure of water during metamorphism is taken to be equal to the load pressure.

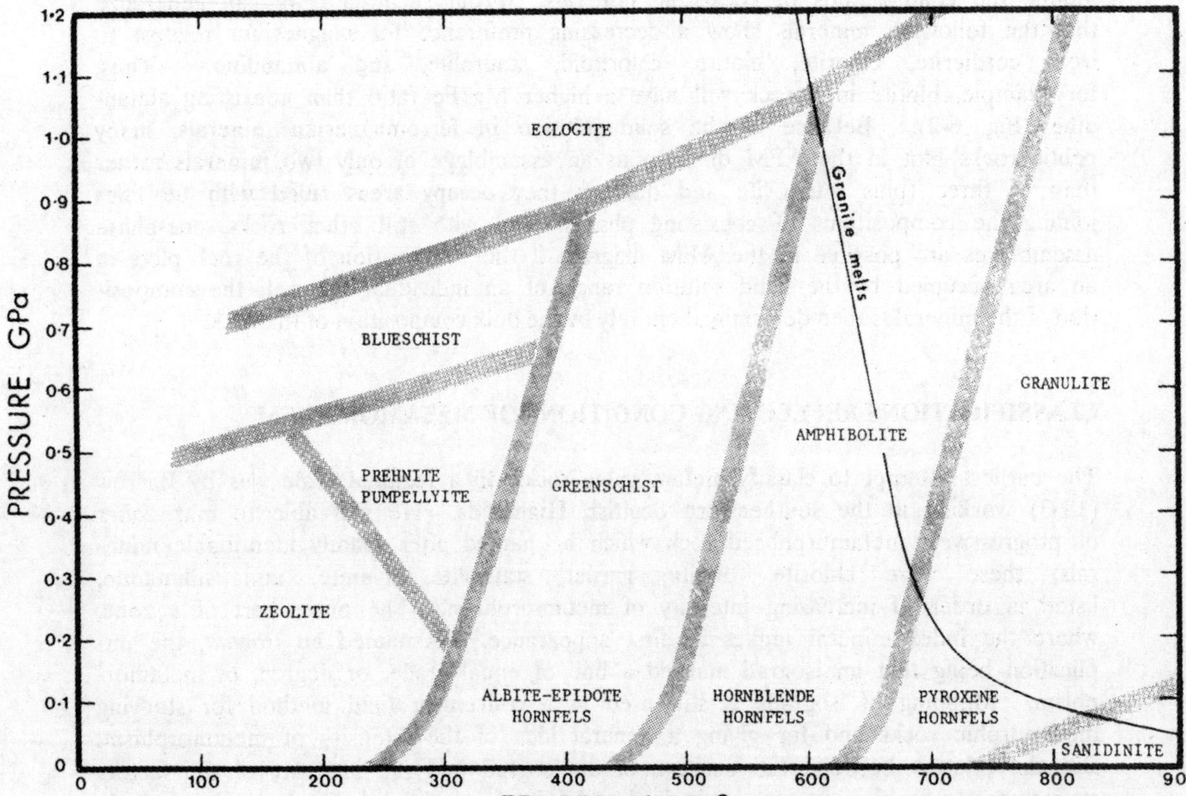

Figure 6-3 Approximate pressures and temperatures under which various metamorphic mineral facies form. Rocks are assumed to be in equilibrium with water at the same pressure as the load pressure. Facies are shifted to lower temperatures if partial pressures of water are less than the load pressure. Boundaries between facies are broad zones in which a number of important reactions take place (Fig. 6-4). The melting curve for peraluminous (S-type) granite under water-saturated conditions is from Clemens and Wall (1981).

The *hornblende hornfels* and the *pyroxene hornfels facies* form at low pressure and moderate-to-high temperatures. These conditions normally occur at the contacts of igneous intrusions. At still higher temperatures and low pressures is the *sanidinite facies*, which is normally restricted to xenoliths in mafic magmas. At the outer margins of some contact metamorphic aureoles, a low-temperature *albite-epidote hornfels facies* occurs. Under conditions of progressive regional metamorphism in orogenic belts, the series of facies *greenschist, amphibolite, and granulite* are normally encountered. In some, a still lower grade, *zeolite facies,* is present, and this may pass up into a *prehnite-pumpellyite facies* before reaching the greenschists. At the low temperatures of the zeolite and prehnite-pumpellyite facies, reaction rates are strongly dependent on deformation and circulating fluid phases. As a result, these two lowest grade facies are not always pre-

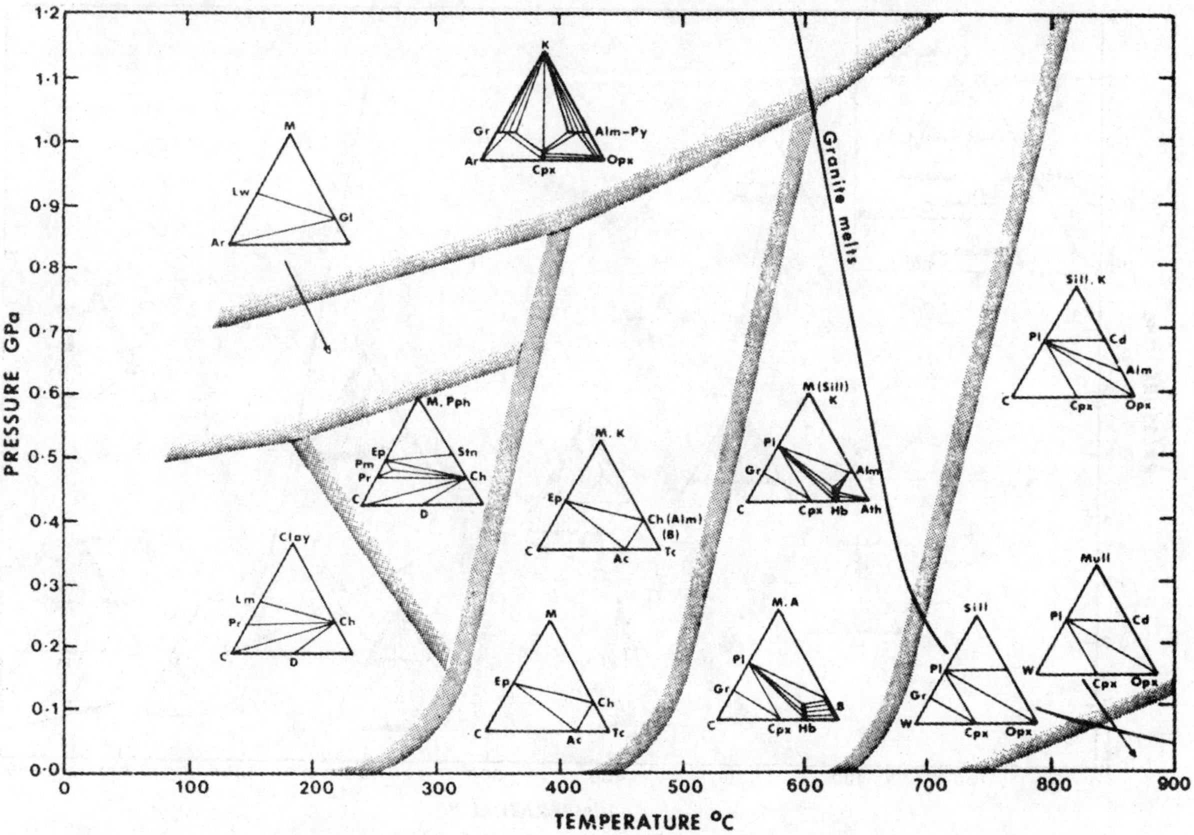

Figure 6-4 ACF plots of common quartz-bearing mineral assemblages in the metamorphic facies. Boundaries and conditions are the same as in Fig. 6-3. Minerals plotted include andalusite (A), kyanite (K), sillimanite (Sill), muscovite (M), mullite (Mull), grossularite (Gr), almandine (Alm), pyrope (Py), orthopyroxene (Opx), clinopyroxene (Cpx), calcic plagioclase (Pl), epidote (Ep), lawsonite (Lw), laumontite (Lm), pumpellyite (Pm), prehnite (Pr), calcite (C), aragonite (Ar), dolomite (D), wollastonite (Wo), actinolite (Ac), hornblende (Hb), glaucophane (Gl), anthophyllite (Ath), talc (Tc), biotite (B), chlorite (Ch), cordierite (Cd), pyrophyllite (Pph), and stilpnomelane (Stn). The beginning of melting curve for water-saturated peraluminous granite is from Clemens and Wall (1981).

sent or recognizable. Instead, greenschist facies rocks commonly appear as the first products of progressive metamorphism. Under conditions of high pressure and low temperature, metamorphism gives rise first to the *glaucophane schist facies (blueschists)* followed by the *eclogite facies.* Such conditions exist only where cold crust is subducted rapidly (hence remaining cool) to great depth and then brought back to the surface rapidly by tectonism.

Each of the metamorphic facies is characterized by a particular mineral assemblage. In meta-igneous rocks, especially those of mafic and intermediate composition and in calcareous rocks, these assemblages are best expressed in terms of the ACF components (Fig. 6-4), whereas in pelitic rocks the AFM projection must be used (Fig. 6-5).

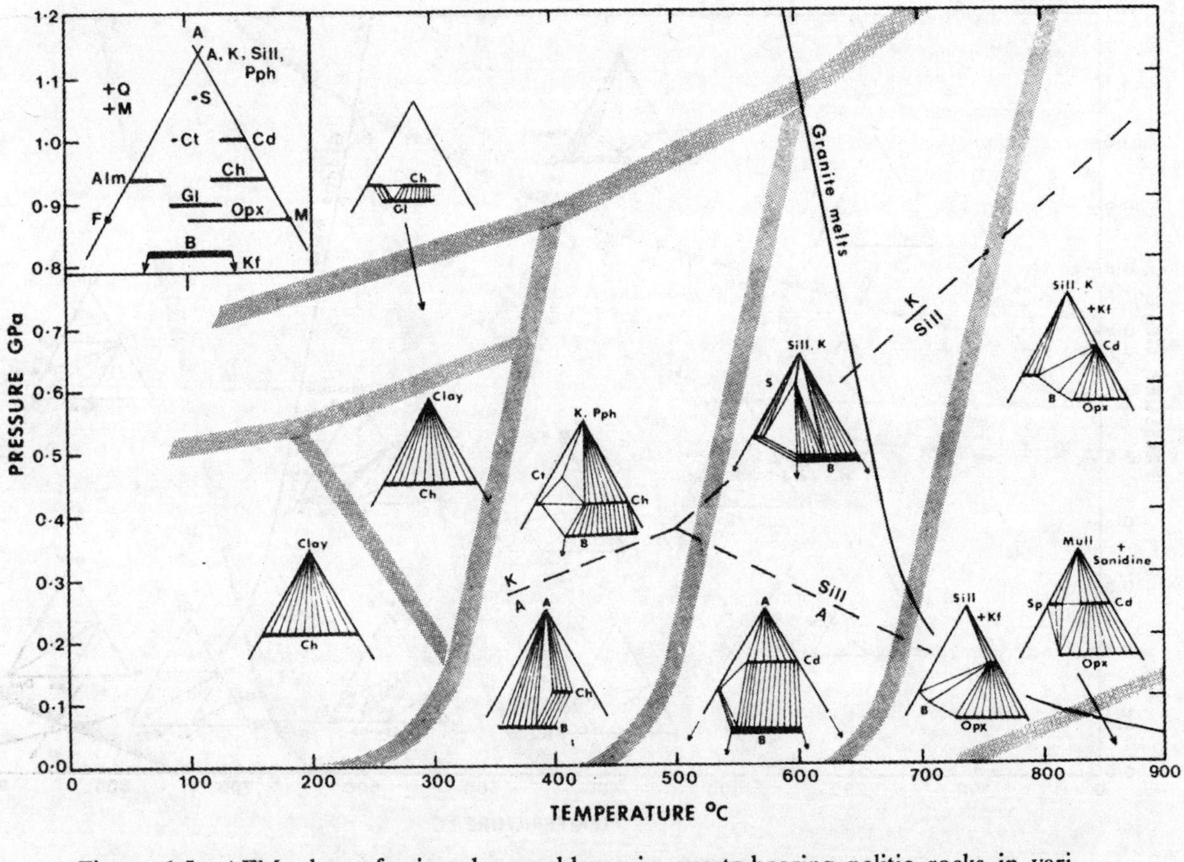

Figure 6-5 AFM plots of mineral assemblages in quartz-bearing pelitic rocks in various metamorphic mineral facies. All conditions and boundaries are the same as in Fig. 6-3. Projections are from muscovite, except in the granulite, pyroxene hornfels, and sanidinite facies where they are from potassium feldspar. Positions of minerals in the AFM projection from muscovite are shown in the inset. Abbreviations are: quartz (Q), muscovite (M), potassium feldspar (Kf), andalusite (A), kyanite (K), sillimanite (Sill), pyrophyllite (Pph), spinel (Sp), almandine (Alm), staurolite (S), chloritoid (Ct), cordierite (Cd), chlorite (Ch), biotite (B), glaucophane (Gl), and ortho- pyroxene (Opx). The Al_2SiO_5 phase relations are from Holdaway (1971).

The boundaries between the various facies are determined by important metamorphic reactions. Inspection of Figs. 6-4 and 6-5, however, reveals that in many cases several reactions must take place to convert the assemblages of one facies into those of the next. This led to the division of the facies into subfacies. But as knowledge of metamorphic rocks grew, the number of reactions that had to be accounted for in this way also grew. It became apparent that subfacies would soon become too numerous to be useful. Nonetheless, there had to be a way of incorporating the large amount of experimental and theoretical data that was becoming available on metamorphic reactions. The solution was to replace the facies concept with what are known as *petrogenetic grids*, which are pressure-temperature diagrams on which all possible reactions for a particular range of rock compositions are plotted. The facies classification is still used, as originally defined, but only for broad regional classifications of rocks.

To illustrate the detailed analysis that can be made of metamorphic mineral assemblages using a petrogenetic grid, Figure 6-6 shows the network of reaction curves that have been worked out for pelitic rocks. Most metamorphic rocks formed at general pressures and temperatures plot within the areas lying between reaction curves on the petrogenetic grid. These rocks contain the same number of minerals as components; that is, on the AFM projection, they consist of three minerals. It will be noted by comparing AFM plots in adjoining fields that only one line is changed in going from one field to the other. This indicates that only one reaction separates the assemblages in adjoining fields. This is in contrast to the facies classification where several reactions may separate adjoining facies.

If the temperature or pressure of metamorphism of a rock falling within a field on the petrogenetic grid were to be increased, there would come a point at which a reaction curve would be encountered. Here the rock would consist of four minerals in the AFM projection, and when these were joined in the diagram, intersecting lines would result. Intersecting lines simply indicate that a reaction is in progress, with the two minerals on one line reacting to form the two on the other. If such a rock were found, you might expect to see textural evidence of reaction. In any case, the temperature and pressure conditions under which the rock formed would be known to lie somewhere along the reaction curve on which the rock plots.

Under very special temperatures and pressures, a rock might plot on the intersection of two reaction curves, which on the AFM plot would result in five coexisting minerals. Such an assemblage is said to be invariant, because neither the temperature nor the pressure could be changed without causing one of the minerals to be lost through reaction. The discovery of such a rock in an area is particularly useful because it defines precisely the temperature and pressure at which that mineral assemblage was formed.

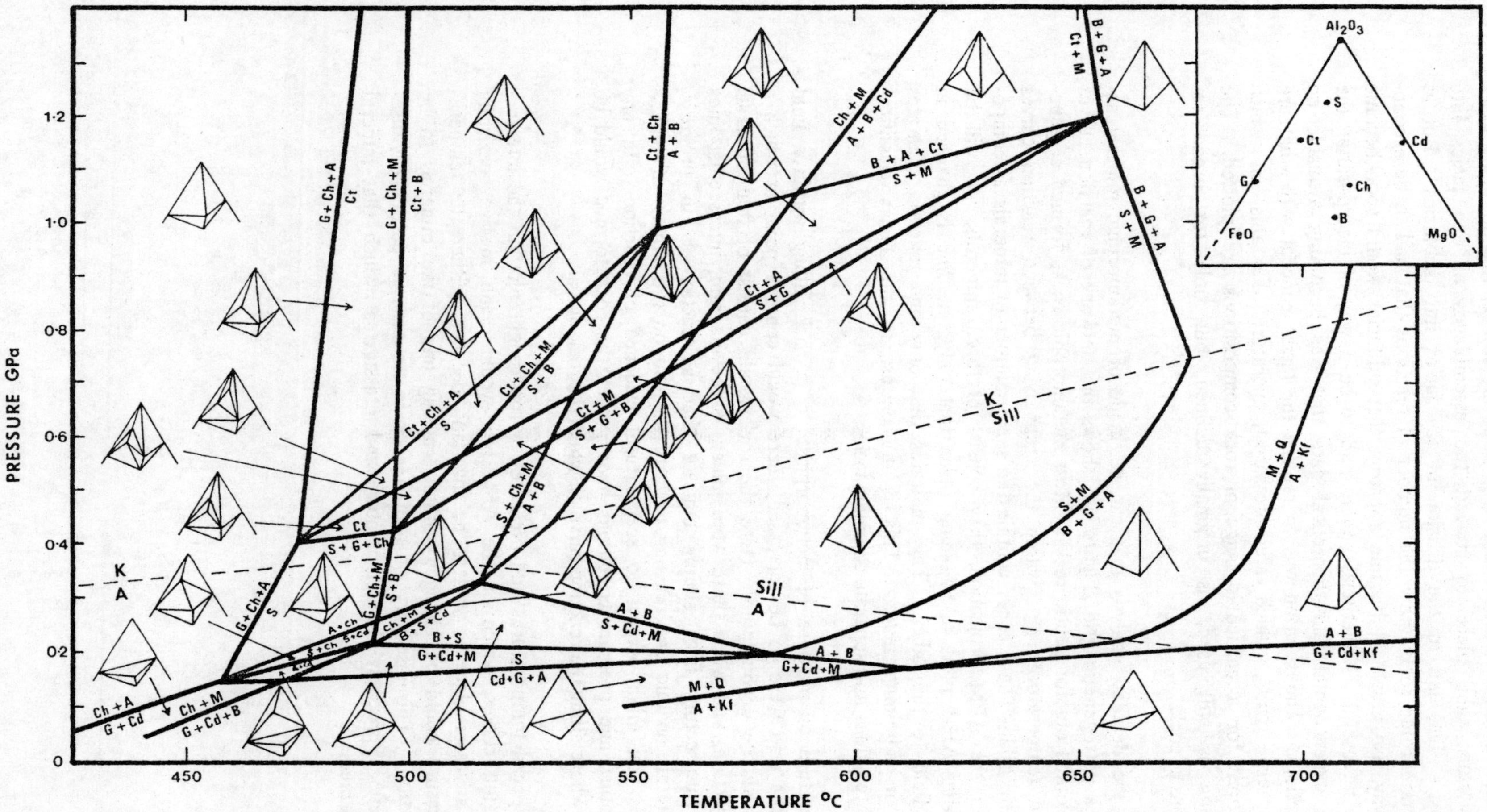

Figure 6–6 Petrogenetic grid for pelitic rocks in the presence of a hydrous fluid phase ($P_{H_2O} = P_{tot}$). The minerals involved are kyanite, andalusite, and sillimanite (A), cordierite (Cd), chlorite (Ch), biotite (B), garnet (G), chloritoid (Ct), and staurolite (S). In addition, all rocks contain quartz (Q) and muscovite (M), or at high temperatures, quartz and potassium feldspar (Kf). Projection into the AFM plot is from muscovite, except at high temperatures where it is from potassium feldspar. Many of the minerals are of variable composition, but in the AFM plot they are shown with fixed compositions for simplicity of plotting (see insert). The compositional variability causes the reactions, which are shown as lines on the grid, to broaden in nature into zones. This increases the possibility of finding rocks containing the reaction assemblages. [Modified from petrogenetic grids by Albee (1965), Hess (1969), and Kepezhinskas and Khlestov (1977). The Al_2SiO_5 phase relations (A, K, Sill) are from Holdaway (1971).]

TEXTURES OF METAMORPHIC ROCKS

Metamorphism normally involves the growth of new grains. These may form through recrystallization of old ones or through the growth of completely new minerals; less commonly, the process simply involves grain breakage. The grain size of the rock depends on the number of nuclei present during crystallization. This number is determined, in a complicated manner, by the degree of deformation of the rock and by the reaction mechanisms through which new minerals form. The textures, which are a product of the manner in which the crystallization takes place, can be conveniently grouped into those formed by recrystallization and those formed by the growth of new minerals. Textures formed by grain breakage can be treated under recrystallization, because most breakage is accompanied by some recrystallization.

Recrystallization Textures

Recrystallization involves the nucleation and growth of strain-free grains from material that has excess energy because it is strained or has excessively large surface area because of fine grain size. Recrystallization allows this energy to be released. Consider,for example, a crystal that has been bent or a compressed coil spring; both have stored strain energy. The energy of the spring can be liberated by releasing a clip and allowing the spring to expand. The rock surrounding a bent crystal, however, may not allow it to return to its original shape, so the grain can release its strain energy only by reorganizing its constituent atoms into a new strain-free grain; that is, by recrystallizing. The compressed coil spring could also have liberated its strain energy by recrystallizing instead of by expanding. The surface of grains also has considerable energy because of distorted or unsatisfied bonds. In very fine-grained rocks the surface area of grains is enormous, and thus the energy that can be released by recrystallizing to a coarser grain size can be very large.

Because most metamorphic rocks have undergone recrystallization, it is important to understand the process. Laboratory study of recrystallization of rocks, however, is difficult because of the slowness of the process. Recrystallization is of great importance to metallurgy, and fortunately metals recrystallize rapidly enough so that their recrystallization can be studied in the laboratory. The results of these studies are directly applicable to metamorphic rocks. Indeed, most metals have textures that are almost identical to some important metamorphic ones.

Many metal products undergo a heat treatment following fabrication which is known as annealing. The metal object is raised to a certain temperature and held there for a prescribed time and then cooled; both the heating and cooling stages may involve several steps. During this process the metal recrystallizes. This releases strain that may have been induced during fabrication, but it also develops textures and grain sizes in the metal which impart desirable properties to the product. The resulting texture, in large part, depends on the amount of strain developed in the metal during fabrication.

When a metal is deformed, for example by passing under a press or through a roller, grains become strained. Some parts, because of crystallographic orientation, juxtaposition of other grains, impurities, or inclusions, develop more strain than others. When the metal is heated, these locations are the first to undergo reorganization into unstrained grains because they have the largest amount of strain energy to release. Nucleation of unstrained grains, therefore, occurs at sites of greatest strain. Once formed, these unstrained grains grow at the expense of the surrounding strained ones. If annealing is carried on long enough, the new unstrained grains continue growing until they encounter other unstrained grains that have grown outward from neighboring nucleation sites.

The grain size of a recrystallized metal clearly depends on the number of nucleation sites formed during annealing, which in turn depends on the degree of deformation prior to annealing. It is found that a certain intensity of deformation is necessary before any sites are formed. Above this level, the number of sites increases with increasing deformation. Thus, a slightly deformed metal may not recrystallize at all and would remain fine-grained. A slightly more deformed metal might develop a few nucleation sites and would, therefore, recrystallize to a coarse-grained metal (MT 1). A still more deformed metal would have more nucleation sites and would consequently be finer-grained after annealing (MT 2).

This same relation between degree of deformation and grain size of the recrystallized product is found in metamorphic rocks. Compare, for example, the grain size of quartzites in figures MT 4 and 5. Both are from the same formation, but flattened and stretched pebbles (MT 3) in both rocks indicate that the sample in MT 5 was deformed much more than that in MT 4. This, then, accounts for the finer grain size of the sample in MT 5. The rock illustrated in MT 15, although of a completely different composition from the quartzite, shows how fine-grained a recrystallized product can be if the deformation is intense. This rock is from a major thrust fault, where deformation induced so much strain that the resulting rock is extremely fine--grained.

The growth of unstrained grains outward from nucleation centers during recrystallization results in polygonal-shaped grains if there are no outside disturbing forces and the crystals do not exhibit marked anisotropy in surface properties or growth rates. Figure MT 5 shows a typical example of the polygonal grains that form by annealing in metamorphic rocks. Note that grain boundaries intersect at angles of approximately 120°. Such a texture is found in both regional and contact metamorphic rocks. In regional ones, it generally indicates that recrystallization postdated deformation; if recrystallization occurs during deformation a planar fabric develops instead. The polygonal pattern is particularly common in contact metamorphic aureoles, for here igneous intrusions provide the heat necessary for annealing without introducing significant directed stresses. Fine-grained contact metamorphic rocks containing polygonal-shaped grains are called *hornfels* (MT 6). Coarser grained rocks with polygonal grains are said to have a *granoblastic* texture (MT 7). The suffix *-blastic* indicates that the texture is of metamorphic origin. When used as a prefix, *blasto-*, it indicates that the texture is inherited from an earlier rock. For example, blastoporphyritic would apply to metamorphosed porphyritic igneous rock. Another contact metamorphic rock that commonly exhibits a granoblastic texture is a *skarn*. This forms from the metamorphism of limestone into which large amounts of Si, Al, Fe, and Mg were introduced from the igneous body to form various calcium-bearing silicates. Skarns are therefore not the product of simple recrystallization, but involve significant changes in bulk composition. When a rock has its composition changed during metamorphism, it is said to have been *metasomatized*. Many metasomatic iron ore deposits are formed as skarns.

Although the deformation necessary to bring about recrystallization may distort original grains or objects, such as fossils or pebbles (MT 3), the recrystallized grains typically exhibit no morphological extension. However, if the period of deformation overlaps with the period of recrystallization, the polygonal grains may take on a *preferred crystallographic orientation*. For example, the equidimensional polygonal grains in many quartzites have a strong preferred orientation. This orientation can be determined by measuring, on a universal stage, the orientation of the optical indicatrix in each quartz grain in a thin section and plotting the results on a stereographic net. The presence of a preferred orientation, however, can be detected on the ordinary flat microscope stage by inserting, for example, the first order red interference filter. If there is a preferred orientation, most quartz grains will show the same interference color, blue or yellow, depending on the orientation of the quartz c axes with respect to the vibration directions in the accessory plate.

During deformation mineral grains may become extremely flattened or stretched, and the effect may be more pronounced on one mineral than on another in the same rock. For example, in many high grade metamorphic quartzo-feldspathic rocks, quartz is typically flattened into extremely thin sheets, whereas feldspar may still retain its original shape or show only slight amounts of strain. The quartz in the sheets, however, completely recrystallizes to one or two large unstrained crystals. Such rocks are known as *flaser gneiss* (MT 14). With increased deformation, the feldspar also begins to undergo recrystallization. This starts with the outer parts of feldspar crystals recrystallizing to very fine aggregates of unstrained grains. Commonly, the core of feldspar grains remains as large strained crystals within the aggregates of fine-grained feldspar and quartz sheets. Because of the eye-like shape of these relict feldspars, this rock is referred to as *augen gneiss* (MT 14).

In zones of intense shearing the original grains of a rock can become greatly stretched or, at high levels in the crust, may even be broken. With such large amounts of strain, recrystallization takes place easily, and because of the large number of nucleation sites, the resulting rock is extremely fine-grained. Such rocks were originally thought to have formed by the grinding or milling of material in the shear zone, and were consequently named *mylonites*. Although some comminution of grains may take place, especially near the surface of the earth, most of the decrease in grain size is now thought to result from recrystallization. Most mylonites, in addition to being extremely fine-grained (also, commonly black or dark), have a prominent lamination formed from the extreme flattening of the grains of the original rock.

In some fault zones mylonitic material is able to intrude into fractures, forming narrow dikes or networks of dikes (MT 16). These are commonly less than 1 cm wide and resemble rapidly quenched diabase dikes. Thin-section inspection, however, reveals their mylonitic origin, for they are seen to contain many small fragments of the surrounding rock embedded in dark, aphanitic groundmass. These are known as *ultramylonite or pseudotachylite* because of their resemblance to volcanic glass (tachylite). At least in some fault zones, strain rates are high enough for frictional fusion to occur, in which case the pseudotachylite is, in fact, composed of glass carrying with it fragments of unmelted rock (MT 17).

Textures Associated with the Growth of New Minerals

During metamorphism, reactions take place to produce new minerals. Unlike recrystallization, which requires modifications of crystal imperfections over distances of possibly only a few atoms, new minerals must make room for themselves by either replacing material (diffusional exchange of material from one place to another) or by physically pushing material aside. Of course, the growth of new minerals is accompanied by the disappearance of others, and this forms part of the diffusional exchange. These mineralogical adjustments, which can take place in the presence or absence of directed stress, give rise to characteristic textures.

Under static conditions, new minerals grow with random orientations, even though they may have platy or needly morphology. They may replace or push aside minerals that formerly occupied their position. If the surrounding minerals have random orientations, evidence for displacement may be difficult to find. Primary sedimentary layering provides convenient reference lines, and when present, is commonly undeflected by the growth of the mineral, indicating that growth occurred through replacement. For example, the cordierite crystal in MT 19 did not push aside the layers on either side of it, although the crystal and the adjoining layers have been slightly rotated.

When minerals form in a stress field, the orientation of that field plays a dominant role in determining the orientation in which the minerals grow. Minerals that form plates, blades, or needles have growth rates that clearly vary with crystallographic direction. Crystal nuclei of such minerals that have their maximum growth rate directions oriented perpendicular to the maximum principal stress will grow more rapidly than those with other orientations, and consequently a preferred orientation will develop with the platy and elongated minerals lying in a plane perpendicular to the maximum compressive stress. In addition, pure shear will tend to rotate unoriented grains into this same plane. The most prominent feature, therefore, of rocks which have undergone metamorphism in a stress field, is a planar fabric which is referred to generally as *foliation*, or as *schistosity* if it is defined mainly by an abundance of micaceous minerals (MT 8), or as *gneissosity* if it is defined by more granular minerals and compositional layering (MT 13).

When fine-grained pelitic sedimentary rocks (shale) undergo regional metamorphism, muscovite and chlorite are amongst the first minerals to form. These grow with a preferred orientation which produces a prominent foliation referred to as *slaty cleavage*. The micaceous minerals in slate are too small to be seen with the unaided eye. As they coarsen, the foliation plane develops a sheen to it and generally becomes slightly less regular. This rock is referred to as *phyllite*. Individual crystals are still too small to be visible to the unaided eye in phyllite. When they do become visible, the rock is called *schist* (MT 8). Schists can also form from mafic igneous rocks; these typically contain abundant chlorite and amphibole. In sedimentary rocks of silty and sandy composition and in intermediate to felsic igneous rocks, fewer micaceous or needly minerals are formed during metamorphism. Those that are present tend to be concentrated into layers that parallel the foliation defined by the micas or amphiboles. Quartz and feldspars tend to concentrate in the alternate layers. Such rocks are referred to as *gneiss*. No hard-and-fast rule separates schists from gneisses. If micaceous or needly minerals predominate, the rock is a schist; if the granular minerals predominate, it is a gneiss. In general, schists tend to pulverize when hit with a geological hammer, whereas gneisses fracture. When needly minerals are present in a schist or gneiss, they may be randomly oriented within the plane of foliation, or they may have a parallel alignment which is known as a *lineation*.

The foliation plane in many schists, rather than being planar, is folded into small crenulations which have a wavelength of a few millimeters to about a centimeter (MT 9). The fold axes define a *crenulation lineation*. These lines actually mark the intersection of a second, more widely-spaced schistosity with the first schistosity. In cross section, the early schistosity is seen to be bent into sigmoid curves by the later *crenulation schistosity* (MT 10, 11). Within the bends the amount of quartz is substantially less than in the surrounding rock, presumably as a result of removal by solution. This results in the crenulation schistosity being marked by mica-rich layers. This new compositional layering must not be mistaken for bedding. Crenulation schistosity can result from successive episodes of metamorphism, but in many cases it is produced by a single episode during which the orientation of the rock changes with respect to the stress field, possibly as a result of folding.

When new minerals are produced during metamorphism, their grain size is determined largely by the number of nuclei formed and to a lesser extent by the abundance of the new phase. When even only small amounts of a mineral are formed, the crystal size can still be large, if the number of nuclei is small. The limiting factor in this case is the distance over which diffusion is able to transport nutrients, which in turn is determined by the duration and temperature of metamorphism and the presence of fluid phases.

The numbers of nuclei formed of each mineral in a rock are unlikely to be identical. Consequently, the grain size of the various minerals will differ. If the range in grain size is significant, the rock is said to be *porphyroblastic*, with the large crystals being termed *porphyroblasts*. Many minerals, such as garnet (MT 18), staurolite, chloritoid, andalusite, and cordierite (MT 19), typically grow as porphyroblasts. It is important to emphasize that the porphyroblastic texture results from crystal growth during metamorphism. If the variation in grain size is inherited from a porphyritic igneous rock, the texture is termed *blastoporphyritic*.

Some metamorphic minerals, especially those that form porphyroblasts (except cordierite and feldspar), develop crystal faces more readily than others. A metamorphic mineral bounded by its own crystal faces is said to be *idioblastic*. The development of crystal faces on a grain depends on the minerals with which it comes in contact. A mineral may be idioblastic in one rock, but in contact with different minerals in another rock, it may be anhedral. By noting the form of the boundary between juxtaposed grains, metamorphic minerals can be arranged according to their ability to form idioblastic grains. This is known as the *crystalloblastic series*. A mineral tends to form idioblastic faces against any mineral below it in the series.

CRYSTALLOBLASTIC SERIES

Magnetite, Rutile, Sphene
Andalusite, Kyanite, Garnet, Staurolite, Tourmaline
Epidote, Zoisite, Forsterite, Lawsonite
Amphibole, Pyroxene, Wollastonite
Chlorite, Talc, Mica, Prehnite, Stilpnomelane
Calcite, Dolomite
Cordierite, Feldspar, Scapolite
Quartz

Many porphyroblasts incorporate inclusions of other minerals during their growth, giving rise to a *poikiloblastic or sieve texture* (MT 23). In some porphyroblasts these inclusions may be concentrated into crystallographic planes, as in andalusite (MM 11). In others, the inclusions may reflect compositional differences inherited from sedimentary bedding or early schistosity. In still others the inclusions may form contorted patterns that indicate movements of the porphyroblast during its growth. If the crystal rotates while growing, inclusions marking bedding or an early foliation can take on a sigmoid shape rather than being planar (MT 21). The inclusion trails in some porphyroblasts may record rotations of more than 180° (MT 22). These contorted layers of inclusions, which are common in garnet and staurolite porphyroblasts, produce a texture known as *helicitic or snowball.* Note that unlike the layering in a rolled snowball which starts in the center and grows outward, the foliation in the helicitic texture can be traced in from one side of the porphyroblast, through the center of the coil, and out the other side. The resulting coil resembles the pattern of a strand of spaghetti that has been wound from its middle onto a fork rather than that of a rolled snowball.

Porphyroblasts in schists are commonly flanked on those faces which truncate the foliation by aggregates of relatively coarse quartz grains, whereas micas tend to be more abundant on the other faces (MT 18). Pressure gradients around porphyroblasts are thought to cause quartz to migrate from regions of higher pressure to ones of lower pressure. Quartz dissolves where the pressure is greatest and precipitates where it is least. A rock may undergo considerable flattening during deformation, and the presence of a hard, relatively undeformable porphyroblast may indeed produce zones of lower pressure on those faces perpendicular to the direction of maximum compressive stress. Consequently, these aggregates are referred to as *pressure shadows*. In some cases, the quartz may nucleate on the surface of the porphyroblast and grow outward as fibers. Because such fibers form a goatee-like extension to porphyroblasts they are referred to as *beards*. If the porphyroblast rotates during growth of the beard, the fibres may form curved crystals.

Detrital grains in deformed sedimentary rocks can also produce local pressure gradients in the rock that results in *pressure solution*. Grains of quartz in a sandstone, for example, will dissolve at those points where the compressive stress forces the grains together; the dissolved quartz then precipitates on the flanks of the grains where the pressure is lower. This is known as *Riecke's principle*. Because the precipitated quartz grows on the grains in crystallographic continuity, a quartzite is formed in which the grains are elongated perpendicular to the maximum compressive stress.

Most minerals in metamorphic rocks have formed through reactions involving other minerals, as is evident from an examination of the mineral assemblages on either side of an isograd. Reaction mechanisms, however, are complex and often involve the exchange of ions through a fluid phase with reaction products nucleating at sites which may bear no spatial relation to the reactants. Most reactions, therefore, do not produce, on the scale of a thin section, simple textural evidence of the minerals that were involved. In some, however, clear reaction rims are formed between minerals, producing a *kelyphitic* or *corona* texture.

Kelyphytic textures are common in metagabbros, where rims of garnet and orthopyroxene (MT 24), or spinel, pargasite, and orthopyroxene (IT 39) separate the original grains of plagioclase and olivine. Reaction products may form concentric monomineralic zones or fine worm-like intergrowths of crystals growing perpendicular to the boundary between the reacting phases. In meta-igneous rocks reaction rims may be of magmatic rather than of metamorphic origin, forming where residual liquid reacted with early- crystallizing minerals. The term corona has been used for those reactions in which magma was involved, and kelyphytic for those involving only solids, but this usage is far from universal. Clear metamorphic reaction rims are seen around many porphyroblasts as a result of retrograde reactions. Garnet, for example, is commonly rimmed by chlorite.

At high temperatures metamorphic rocks can undergo melting. Fusion first takes place at boundaries between minerals that provide the components necessary to form low-temperature eutectic melts. In most rocks this involves quartz and feldspar (MT 25). If this partial fusion takes place at shallow depth, for example at the margin of a wide diabase dike or in a xenolith in a near-surface mafic magma, cooling will likely be sufficiently rapid to preserve the melt as a fine-grained granophyric intergrowth along the boundaries between quartz and feldspar grains (MT 26). In some cases, sufficient melt may be formed in a contact metamorphic rock to make the country rock mobile enough to intrude back into the mafic magma that was responsible for the fusion in the first place. This melting of country rock is termed *rheomorphism*.

Partial melting of metamorphic rocks also takes place during regional metamorphism. In Figs. 6-4 and 6-5 the temperatures and pressures in the upper amphibolite facies can be seen to be above the minimum melting curve for water-saturated granite. Rocks containing quartz and alkali feldspar will, therefore, undergo partial fusion at these temperatures and pressures in the presence of a water-rich fluid phase. The amount of melt formed depends on what fraction of the metamorphic rock is of granitic composition. The non granitic fraction is more refractory and remains in the solid state. The resulting mixture of igneous and metamorphic rock is termed *migmatite*. The igneous fraction is almost always of granitic composition and light colored, whereas the metamorphic one is mafic and dark colored (MT 27). Migmatites commonly exhibit evidence of flow, with the metamorphic fragments being pulled apart and even rotated in the granitic melt. Melting can be carried to the stage that fragments of metamorphic rock become indistinct wisps of concentrations of mafic minerals in the granitic host. Much of the transformation of the metamorphic rock at this stage, however, is carried out by metasomatism rather than simple melting.

REFERENCES

Albee, A. L., 1965, A petrogenetic grid for the Fe-Mg silicates of pelitic schists: Amer. J. Sci., v. 263, p. 512-536.

Barrow, G., 1893, On an intrusion of muscovite-biotite gneiss in the southeast Highlands of Scotland, and its accompanying metamorphism: Quart. J. Geol. Soc. London, v. 49, p. 330-358.

Clemens, J. D. and Wall, V. J., 1981, Origin and crystallization of some peraluminous (S-type) granitic magmas: Canadian Mineralogist, v. 19, p. 111-131.

Eskola, P., 1920, The mineral facies of rocks: Norsk Geol. Tidsskr., v. 6, p. 143-194.

Hess, P. C. 1969, The metamorphic paragenesis of cordierite in pelitic rocks: Contr. Mineral. Petrol., v. 24, p. 191-207.

Holdaway, M. J., 1971, Stability of andalusite and the aluminum silicate phase diagram: Amer. J. Sci., v. 271, p. 97-131.

Kepezhinskas, K. B. and Khlestov, V. V., 1977, The petrogenetic grid and subfacies for middle-temperature metapelites: J. Petrol., v. 18, p. 114-143.

Thompson, J. B., 1957, The graphical analysis of mineral assemblages in pelitic schists: Amer. Mineralogist, v. 42, p. 842-858.

METAMORPHIC TEXTURES

RECRYSTALLIZATION

1. Recrystallized aluminum metal after being rolled and then annealed at high temperature. Originally this metal was extremely fine-grained. It was deformed by being passed through a roller. This induced sufficient strain in the crystals to bring about recrystallization when the metal was heated. Nucleation of new grains, however, occurred only at the few points where the strain was exceptionally high. As a result the recrystallized aluminum is coarse-grained. The slight grain elongation is in the direction the metal was passed through the roller. Reflected light, X10.

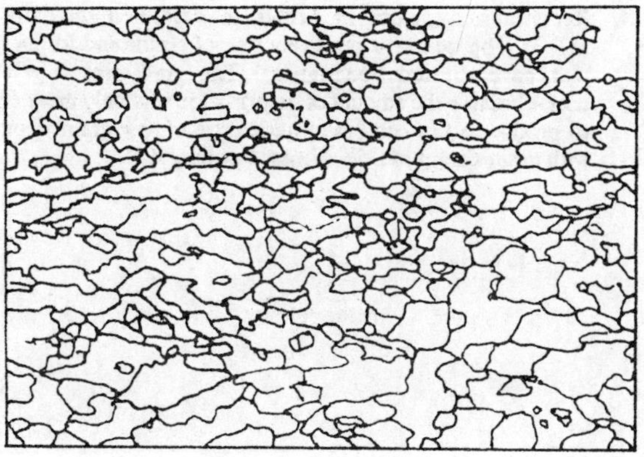

2. Fine-grained recrystallized aluminum metal. This piece of aluminum was treated in the same way as that in MT 1 except that the space between the rollers was reduced to one half. This caused greater deformation, and many more centers of nucleation of new grains were formed. The annealed metal is therefore much finer-grained than is the sample in MT 1. Reflected light, X10.

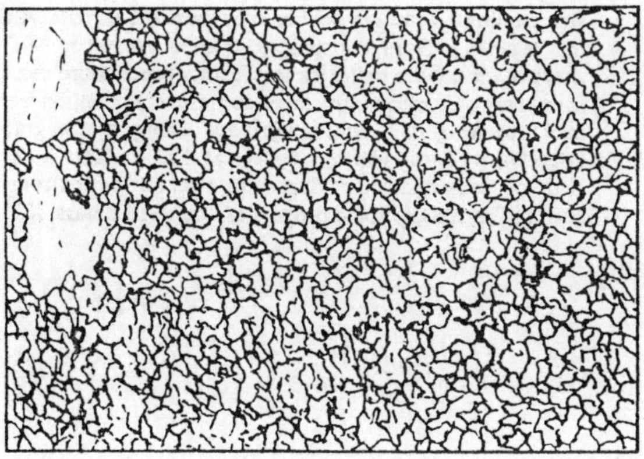

3. Elongated and flattened pebble of quartz in the Silurian Clough quartzite of Connecticut. This pebble, which was originally probably almost spherical, now has a maximum to minimum dimensional ratio of 10 to 1. This deformation introduced sufficient strain to form nucleation centers for the growth of new unstrained grains (see MT 4).

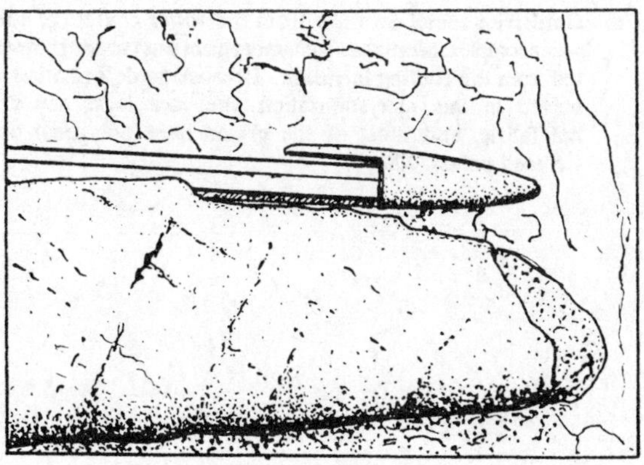

METAMORPHIC TEXTURES

RECRYSTALLIZATION

4. Recrystallized quartz in pebble in deformed quartzite (see MT 3). The quartz is relatively free of strain and forms grains that have sutured boundaries. The grain siz: is relatively coarse because the amount of deformation was only great enough to produce a few centers of nucleation of new grains (compare with grain size in MT 5). Crossed polars, X20.

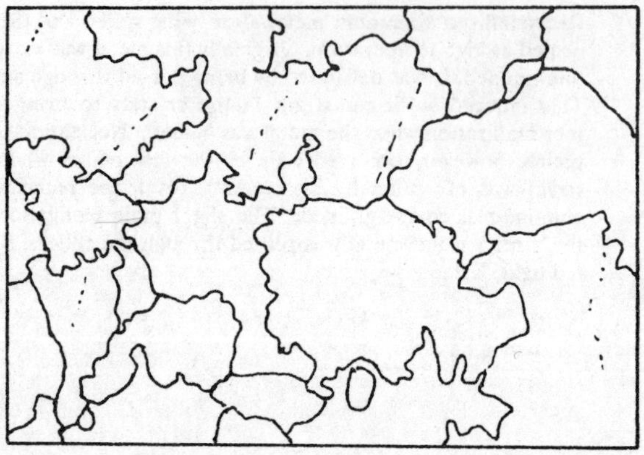

5. Recrystallized quartz in pebble in deformed quartzite. Although this sample is from the same formation as that in MT 4, the shape of the pebble indicates that it underwent more intense deformation. Consequently, more centers of nucleation were formed, resulting in a finer grain size. Many of the grains are bounded by faces which intersect at 120°, which is characteristic of well-annealed materials that do not have strongly anisotropic crystallographic properties. Crossed polars, X20.

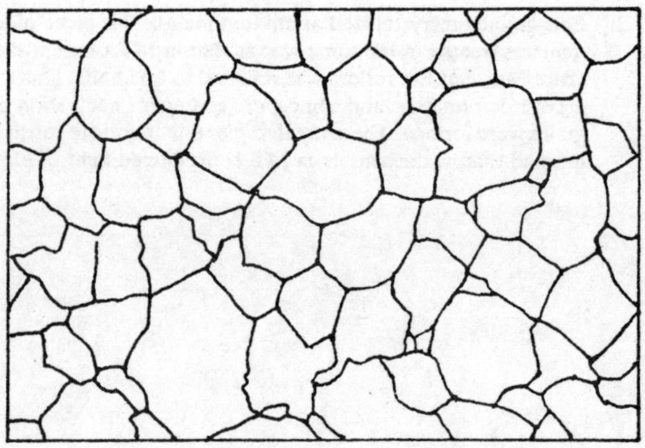

6. Cordierite-spinel hornfels from the lower contact of the Stillwater complex, Montana. Metamorphism was caused by heat liberated from the cooling intrusion. Because no deformation was involved in this recrystallization, the rock lacks any directional fabric, and most of the grains have polygonal outlines. Crossed polars, X8.

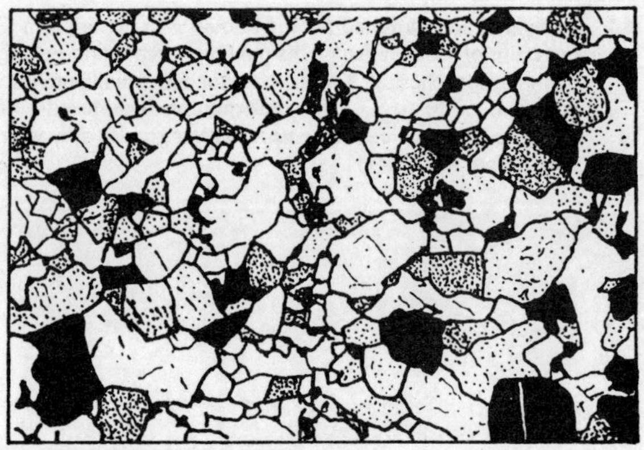

METAMORPHIC TEXTURES

RECRYSTALLIZATION

7. Pargasitic amphibole exhibiting a well-developed granoblastic texture with typical 120° grain boundary intersections. Amphiboles in many metamorphic rocks form elongated prisms (MT 12, MM 33, 48, 49). The fact that grains in this sample from the Adirondacks, New York, are equant-shaped is clear evidence that annealing occurred after deformation had ceased. Crossed polars, X30.

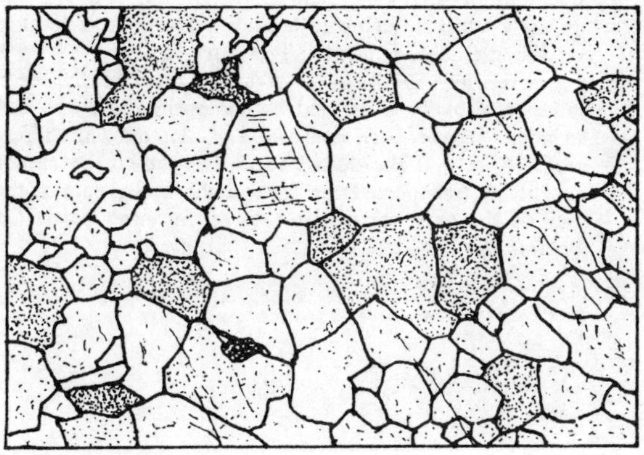

8. Muscovite-biotite-chlorite schist. Crystallization of platy minerals under a directed stress results in a preferred orientation of the plates normal to the direction of maximum stress. The resulting foliation is one of the most prominent features of regionally metamorphosed rocks, especially those of pelitic composition. Although muscovite and biotite in this sample are oriented parallel to the foliation, the chlorite crystals (very low interference colors--dark) cut across the foliation and presumably formed at a late stage when there was no directed stress. Crossed polars, X8.

9. Crenulation lineation (parallel to the pen) formed by the folding of an earlier schistosity. The crenulations actually mark the intersection of a younger schistosity with the older schistosity (MT 10). Unlike the first schistosity, which involves all of the platy minerals in the rock, the later crenulation schistosity is developed on planes that are spaced millimeters apart.

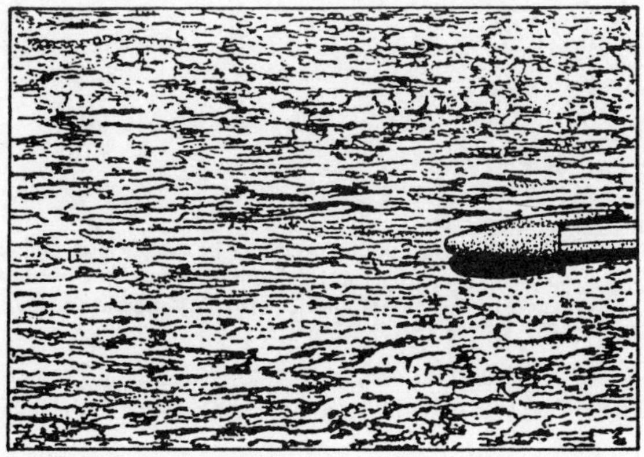

METAMORPHIC TEXTURES

RECRYSTALLIZATION

10. Crenulation schistosity (left to right) intersects and deforms an early schistosity (top to bottom). During the development of the crenulation schistosity, quartz is removed from the zones in which deformation occurs. Consequently, the crenulation schistosity is defined by compositional zones enriched in mica. This is a clear example of metamorphic differentiation. Note that the deformation that caused the crenulation schistosity transported the rock in the upper part of the section to the right relative to that in the lower part. Crossed polars, X8.

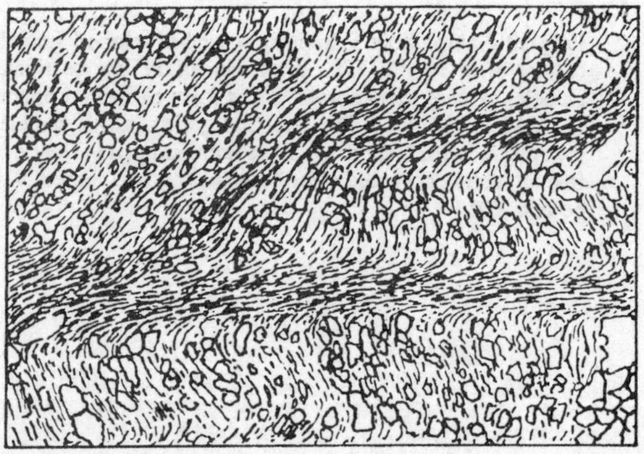

11. Crenulation schistosity parallel to the axial plane of a fold. A lineation formed from the intersection of the crenulation schistosity with the earlier schistosity and compositional layering parallels the axis of the fold. On the cross-section of the fold, the relative motion across individual crenulation schistosity planes is seen to be different on either side of the fold; to the left, the left side moved down, and to the right, the right side moved down. A small sample from anywhere on this fold would, therefore, provide evidence of the attitude of the fold axis and which side of the major fold the sample came from.

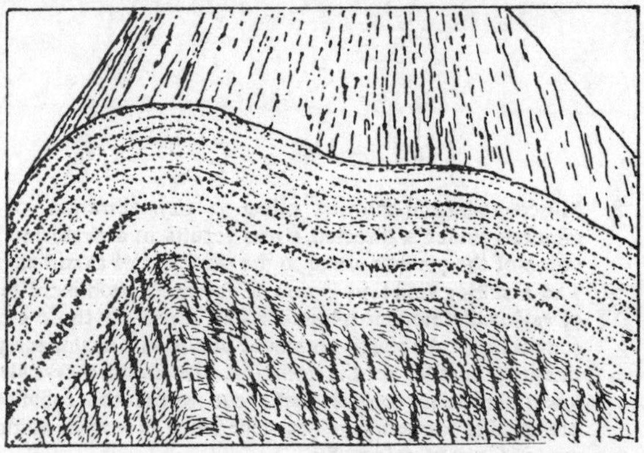

12. Amphibolite composed of plagioclase, hornblende (deep green), and cummingtonite (pale). The schistosity in this rock is defined by the rough alignment of elongate amphibole crystals. Plane light, X20.

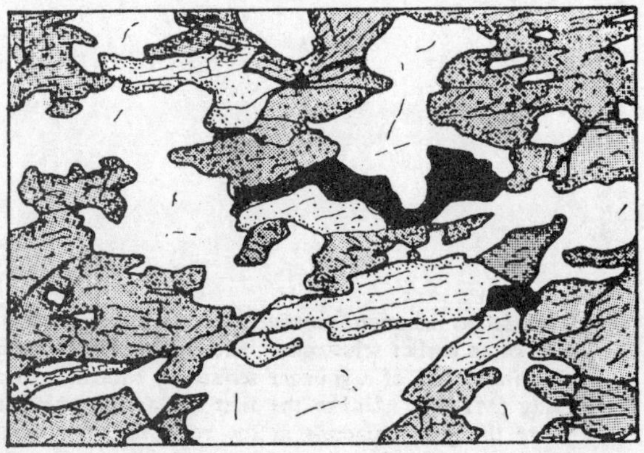

13. Quartz-feldspar-biotite gneiss. The foliation is defined, not only by the parallelism of biotite crystals, but by the segregation of the quartz and feldspar into separate layers. Plane light, X5.

METAMORPHIC TEXTURES

DEFORMATION

14. Augen gneiss, consisting of perthitic feldspar augen (eyes) surrounded by fine-grained feldspar and lenses of quartz. This rock underwent considerable deformation, but its two constituent minerals reacted differently to the deformation. The feldspar augen exhibit considerable strain--wavy extinction and folded exsolution lamellae. The small grains of feldspar may be relicts of larger grains or the products of recrystallization. The quartz forms flattened lenses, which, under crossed polars, are seen to consist of a relatively few, large, unstrained grains. Clearly the quartz is completely annealed, but the feldspar in the augen is not. Crossed polars, X8.

15. Mylonite from the Honey Hill fault, Connecticut. In fault zones the degree of deformation may be sufficiently intense that so many recrystallization nucleation sites are formed that the resulting rock is extremely fine-grained. The name mylonite (mill) is used for these rocks because their fine grain size was thought to result from grinding or milling of rock on the fault surface. Although some comminution may occur in this way, the grain size is largely the result of recrystallization in response to high strains. The mylonitized rock in this sample is a meta-gabbro consisting of plagioclase and amphibole. Plane light, X8.

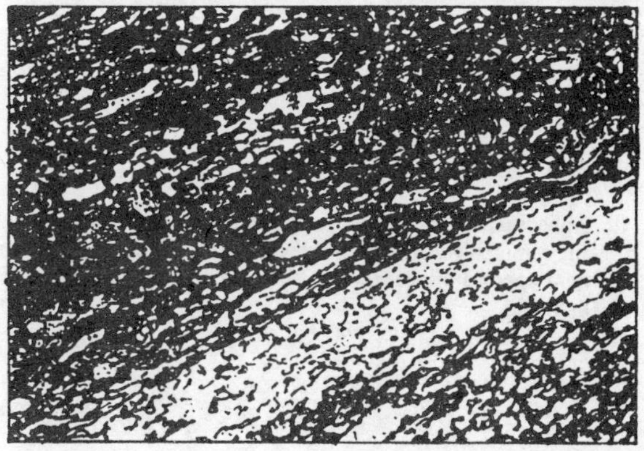

16. Pseudotachylite veining granite, near a major fault in Grenville-age rocks near the southern margin of the Canadian Shield, northeast of Montreal, Quebec. Along some faults and around meteorite explosion craters, the degree of deformation is so intense that the rocks are frictionally fused. The resulting liquid is injected into nearby fractures where it quenches to a glass or extremely fine-grained flinty-appearing rock. It is named pseudotachylite because of its resemblance to dark volcanic glass (tachylite).

17. Glassy vein of pseudotachylite intruded into mylonite in granite, from the same locality as the exposure in MT 16. The glass contains many rounded and partially melted fragments of quartz and feldspar. The glass at the margin of the vein is dark due to devitrification. Plane light, X8.

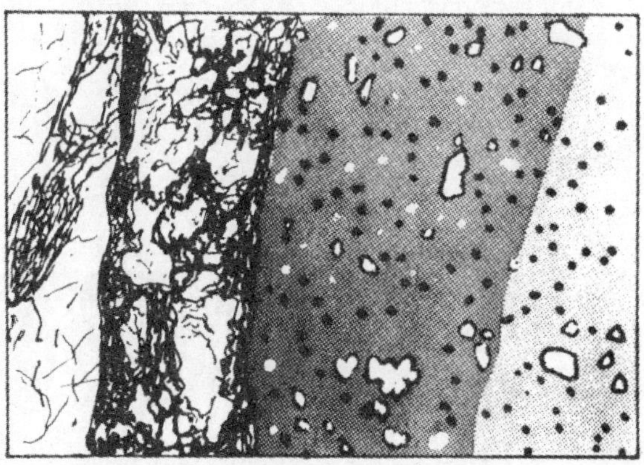

METAMORPHIC TEXTURES

GROWTH OF NEW MINERALS

18. Euhedral porphyroblast of garnet in quartz-muscovite schist. The grain size of a metamorphic mineral depends on the abundance of nucleation centers present during crystallization. Porphyroblasts form from those minerals that have few nucleation centers. Porphyroblasts may or may not be euhedral depending on their power of crystal growth relative to that of the surrounding minerals. Garnet porphyroblasts are commonly euhedral. Plane light, X8.

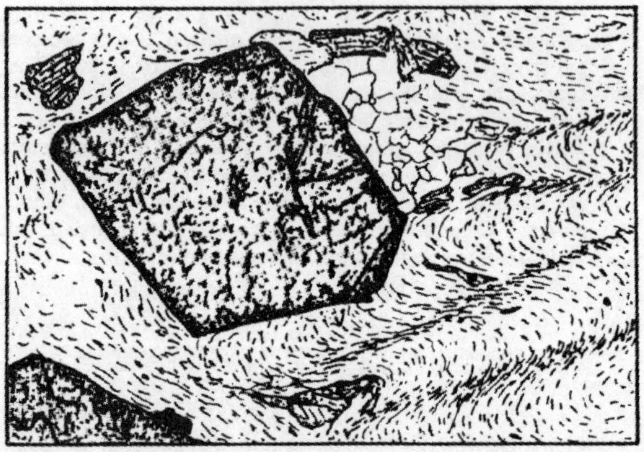

19. Anhedral porphyroblast of cordierite. Porphyroblasts of cordierite commonly form with this ovoid shape, which imparts a prominent spotted appearance to the rock in hand specimen. Plane light, X8.

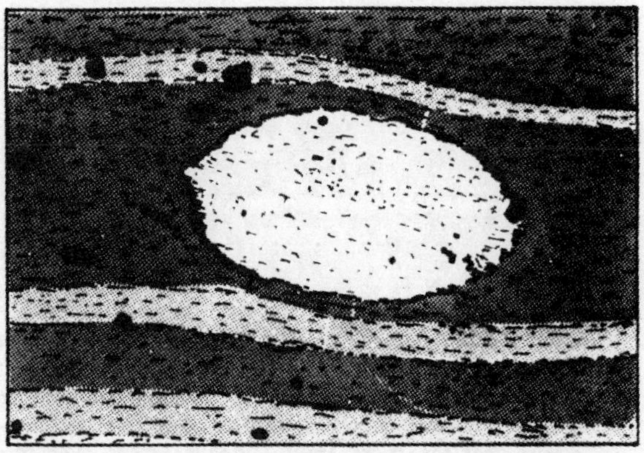

20. Porphyroblast of microcline in amphibolite-grade gneiss. Microcline porphyroblasts can be very large and verge on being pegmatitic. In some cases, their formation may involve a volatile-rich melt formed by partial melting.

METAMORPHIC TEXTURES

ROTATION

21. Porphyroblast of garnet enclosing layers of quartz which are relicts of the original foliation. The foliation plane preserved by the quartz layers in the garnet are at an angle of approximately 30° to the foliation in the surrounding host rock. This indicates that the garnet crystal has rotated counterclockwise by this amount since it formed. Plane light, X5.

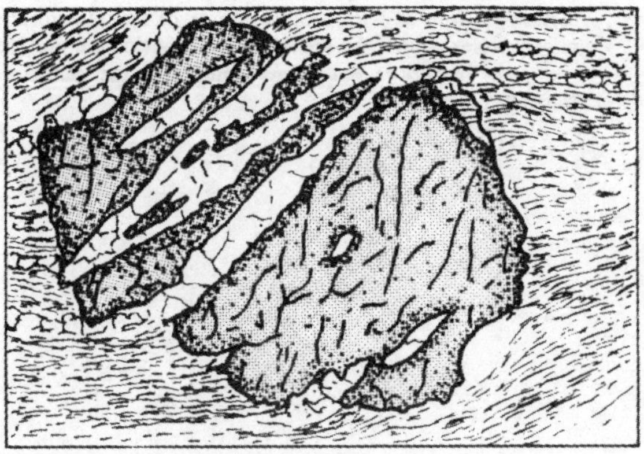

22. Rotated garnet porphyroblast (helicitic or snowball texture). Foliation planes preserved as layers of quartz in this garnet porphyroblast have rotated 180° in a clockwise direction relative to the foliation in the enclosing rock. Note that the quartz layers in the garnet crystal are curved, indicating that the garnet was growing at the same time it was rotating. The relatively straight layers in the core of the crystal, however, indicate that most of the rotation occurred during the growth of the outer part of the crystal. Plane light, X5.

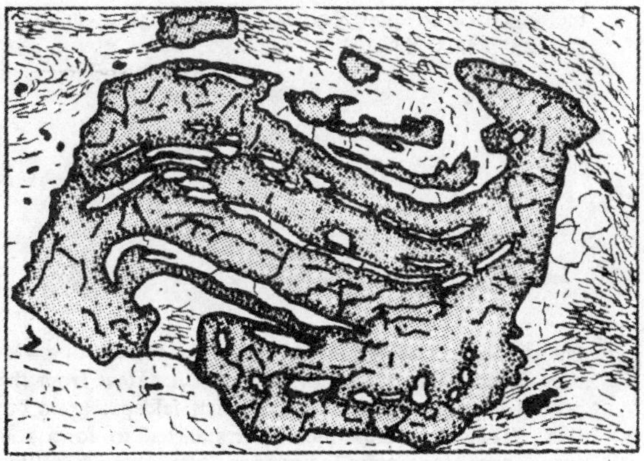

23. Poikiloblastic texture. Some porphyroblasts contain many inclusions of other minerals, specially those not involved in the formation of the porphyroblast. In this sample, porhyroblasts of tourmaline enclose many grains of quartz. Crossed polars, X8.

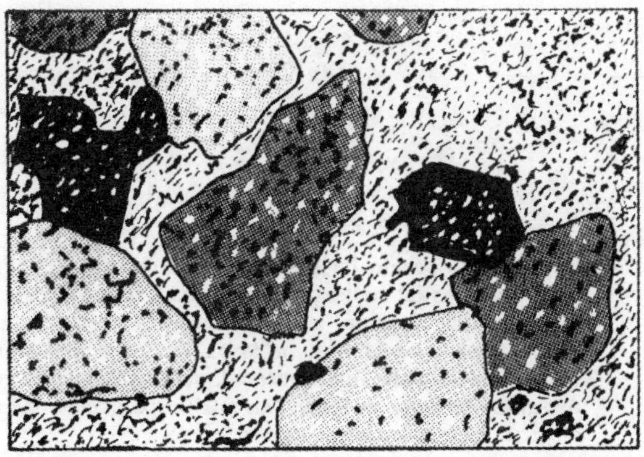

METAMORPHIC TEXTURES

24. Corona texture in metamorphosed olivine gabbro from the Adirondacks, New York. Olivine and plagioclase reacted together to form a zone of garnet and orthopyroxene. In addition, the plagioclase is clouded with minute grains of spinel, except near grain margins where the spinel diffused out of the plagioclase. One grain of biotite has been engulfed in the reaction rim. Plane light, X8.

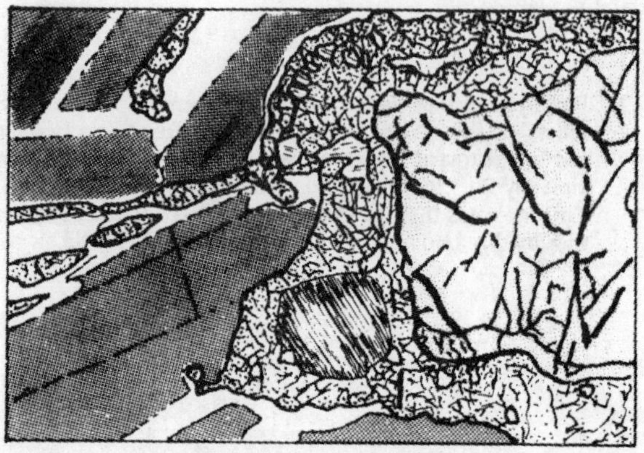

25. Quartzo-feldspathic gneiss that was partially melted in the laboratory and then rapidly quenched so that the liquid was preserved as a glass (isotropic). Melt formed mainly along grain boundaries where quartz and feldspar were in contact, because here the two minerals caused mutual lowering of their melting points (see IT 19). Where quartz grains only are in contact (two clusters near center of section), there was no melting because quartz alone had too high a melting point. Crossed polars, X30.

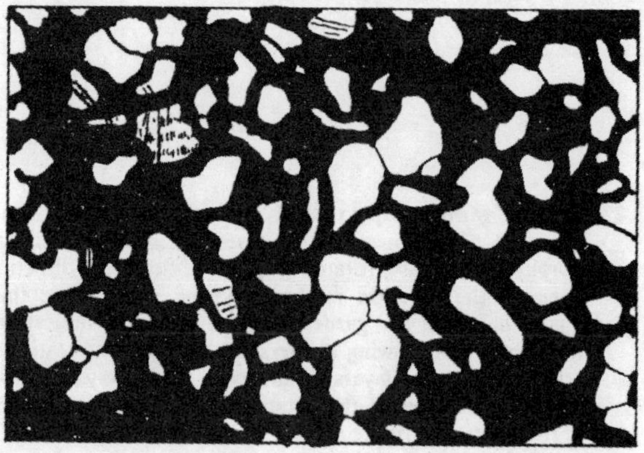

26. Quartz-plagioclase gneiss from near the contact of a wide diabase feeder dike in eastern Connecticut. Heat from the dike caused melting to occur at the quartz-feldspar grain boundaries. This melt subsequently crystallized to form zones of granophyre along grain boundaries. Note that no melting took place at quartz-quartz or feldspar-feldspar grain boundaries. Only where quartz and feldspar were in contact was the low melting eutectic mixture present. Crossed polars, X30.

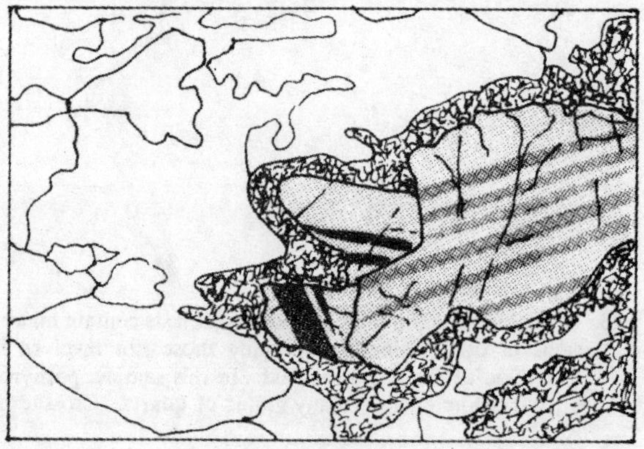

27. Migmatite, Hopedale, Labrador. At the highest grades of metamorphism, partial melting may occur on a regional scale. The lowest-melting fraction, which is usually of granitic composition, forms a liquid host in which the refractory residue--typically more mafic--is fragmented, pulled apart, and streaked out in the direction of flow. This mixture of igneous and metamorphic rocks is known as migmatite.

Index

ROCK NAME:

Number

LOCATION:

HAND SPECIMEN DESCRIPTION:

Thin Section Sketch

MINERALOGY:	Name	Modal %	Optical Properties	Composition
Essential:				
Accessory:				

TEXTURES:

PETROGENESIS: